FARM BUSINESS MANAGEMENT

FARM BUSINESS MANAGEMENT

The Human Factor

Peter L. Nuthall

www.cabi.org

CABI is a trading name of CAB International

CABI Head Office	CABI North American Office
Nosworthy Way	875 Massachusetts Avenue
Wallingford	7th Floor
Oxfordshire OX10 8DE	Cambridge, MA 02139
UK	USA

Tel: +44 (0)1491 832111	Tel: +1 617 395 4056
Fax: +44 (0)1491 833508	Fax: +1 617 354 6875
E-mail: cabi@cabi.org	E-mail: cabi-nao@cabi.org
Website: www.cabi.org	

A catalogue record for this book is available from the British Library, London, UK.

Library of Congress Cataloging-in-Publication Data

Nuthall, P.L. (Peter Leslie)
    Farm business management : the human factor / Peter L. Nuthall.
        p. cm.
    Includes bibliographical references and index.
    ISBN 978-1-84593-598-6 (alk. paper)
1. Farm management--Decision making. 2. Farm managers. I. Title.

S561.N87 2010
630.68--dc22
                                2009023326

ISBN-13: 978 1 84593 598 6

Typeset by SPi, Pondicherry, India.
Printed and bound in the UK by CPI Antony Rowe Ltd.

Contents

The Author

Peter Nuthall has spent many years teaching and researching aspects of farm management. In addition he has developed and managed a team involved in producing and supporting computer-aided management systems used by large numbers of farmers. While most of his time has been at Lincoln University in Canterbury, New Zealand, he has also researched and/or taught at the University of Queensland, Purdue University (Indiana), the University of Kent, and the University of Edinburgh. He has also worked for the UK Milk Marketing Board while based at the University of Nottingham (Sutton Bonnington). Many institutions involved in researching and teaching farm management have been visited to gather ideas. These include the Royal Agricultural College (Cirencester), Gatton Agricultural College (Queensland), Swedish University of Agriculture Sciences (Uppsala), Wageningen University (The Netherlands), University of Guelph (Ontario), TEGASAC (Dublin), MAFF (UK), Cambridge University, Systems Support Centre, Denmark (Arhus), Texas A&M University, and similar. He also has experience of agriculture in a diverse range of situations including Russia, India, Fiji, Australia, New Zealand and the UK. Nuthall has published widely in scientific journals throughout the world, and in monographs as well as having items taken up by the popular farming press.

Acknowledgements

Many people have contributed to the material in this package. They include researchers around the world, farmers similarly, consultants and textbook authors. Gratitude and thanks are offered to all these people.

Above all, of special value was the contribution made by the case study farmers and consultants who agreed to give full accounts of their lives and views on many issues. The true names have not been provided as some of the material is quite personal. Special gratitude is offered to these people.

In addition, the professionals at CABI have been a source of careful assistance and reminders. These have ensured the manuscript was checked with considerable care, and provided in a reasonably timely manner. But above all, Sarah Hulbert, the Commissioning Editor, has shown faith in the concept, and valuable ideas for improvements. To all these people I offer a heartfelt thank you.

The publishers and authors who have provided permission to reprint some of their work are all gratefully acknowledged. The body of researchers and writers worldwide rely on each other for advancement and have a special bond. In particular

- Elsevier through Rightslink have provided permission to use some of the tables in Chapter 5 (Tables 5.1, 5.4, 5.5 and 5.10) and some of the commentary to these tables. This material was produced by the author of this book.
- The Editor (Janice Osborn) of WAERSA, published by CABI, has provided permission to reprint Table 1.2 (Human factors in farm management) authored by Muggen G. (1969).
- Blackwell Publishing for Figs 3.2 and 3.4 that appear in the *Australian Journal of Agricultural and Resource Economics* 53, 413–436 in an article titled 'Modelling the origins of managerial ability in agricultural production', written by the author of this book.

- Elsevier through Rightslink have provided permission to use the decision process table which appears in Section 4.2.4 (created by Ohlmer *et al.* (1998) in the article in farmers' decision-making processes which appeared in the *Journal of Agricultural Economics* 18, 273–290 titled 'Understanding farmers' decision making processes and improving managerial assistance'.)
- The HayGroup (Ginny Flynn, publisher) for permission to reprint details of the Kolb learning-style types, which appears in Section 2.8. Source: *Kolb Learning Style Inventory* © Experience Based Learning Systems, Inc. All rights reserved. The content appears here with the specific permission of its publisher, Hay Group, Inc., 116 Huntington Ave. Boston, MA 02116.

Finally, I am most appreciative of David Hollander's efforts to find suitable photographs to reinforce some of the messages. David, the photographer at Lincoln University, contributed seven of the photos.

List of Figures

List of Tables

1 Introduction

General Overview

Most production economists refer to the production factors as land, labour and capital. While 'labour' might embody the managerial decision-making input as well as physical labour, it is clearer to separate management as a fourth factor of production. The decisions on how to use the production inputs and resources, and the implementation of the plans, are the responsibility of this fourth factor – management. In that the quality of the decisions gives rise to the success of the operation, this managerial skill is clearly absolutely critical to efficiency and profit. However, no texts and courses include the management factor in any depth. This book sets this situation to rights.

Texts on production economics cover the optimal allocation of resources. However, they largely assume that man is a rational being with near-perfect information. The reality is quite different. Managers are human. This means they react in an emotion-determined way. People observe the world around them and come to a conclusion about the current situation. Their mind, perhaps with the aid of calculations, comes to a decision over what actions should be taken. Thus, cues are observed that trigger action, or possibly inaction in some situations. This observation–decision–action process is something that varies with different individuals, and needs to be understood if a farm manager is to improve the decisions aimed at achieving the farm's objectives.

The purpose of this book is to outline the human components of what makes a person, and why a manager acts in particular ways. This understanding is essential in assisting farmers to improve their management, and thus attain their objectives. This assumes that improvement is indeed possible using various techniques. Evidence pointing in this direction will be reviewed. As understanding provides wisdom, the emphasis is on looking at theories and their application in contrast to rote-learning rules and solutions. These seldom cover the myriad of situations possible, thus leading to misinterpretations and mistakes. Armed with knowledge on the parameters that determine how an individual reacts, and their relationships, it is then possible to treat each unique case on its merits.

There is ample evidence that farm managers vary markedly in their skills. Profit and efficiency data from farms in similar environments make this clear. For example, studies that relate individual farms' position relative to their peers show the average technical efficiency can be as low as 36% (a Swedish study) and as high as 85% (a Pakistani case). Of course, these studies assume that all the farmers have the same objective. In reality, some will be happy to be less technically efficient if it means, for example, their average production is quite stable. Similar ranges in efficiency exist when using economic outcomes as the yard stick. One study for Brazilian farmers put their average efficiency at 13%, in contrast to a US dairy study giving an average efficiency of 70% (Dhungana, 2000).

These efficiency studies determine which farmers are producing the greatest output for given inputs, and then relate other farmers against these producers as benchmarks. If the average efficiency was 100%, this would mean all farmers are producing the same ratio of output to inputs. As a comparative measure, this does not necessarily mean that the 'efficient' farmers are in fact efficient in an absolute sense. No doubt they could increase their efficiency through even better decision making, using the latest technologies. This can only be judged if, for example, comparisons with perfectly managed demonstration farms are possible.

The farmers that are the most efficient in a sample can be called 'experts'. Studies of experts show they have particular attributes. These include:

- their expertise is restricted to a limited domain of operation;
- good at clearly defining a problem;
- accurately observe relevant cues and the importance of each;
- automatically perceive meaningful patterns;
- come up with solutions almost instantly;
- having superior short- and long-term memory;
- observing and characterizing a problem in terms of its basic structure;
- spending considerable time quantifying and analysing problems, particularly if not faced before, thus adding to their understanding and store of knowledge;
- clear self-monitoring abilities leading to improvement (i.e. good and objective self-criticism).

Relative to novices, experts know what to observe, do so quickly and accurately, and use their experience to provide a solution that is appropriate. If the problem has never been experienced before, their superior processing systems work out the solution, and then store this for future reference. With time, they become real masters. If any one of the characteristics of an expert is not present, the person will not attain the true expert classification, though there will be degrees of expertise. The important questions relate to the characteristics required to become an expert, and how this status can be acquired.

It was pointed out experts are quick with their judgement and decision. This is most likely due to pattern matching. This refers to having the pattern of a problem, and its solution, stored in memory so that when the expert observes the values of the critical parameters, this set of data is sent to memory to find a match with the stored information. When found, the solution is readily available. These patterns might be visual (a picture of some kind), or abstract (lists of benchmark data perhaps). For example, if you see a bush with flowers on it of a certain nature, if you have seen the bush before and its image is stored in your memory, it is immediately recognized as, say, a rhododendron. Analysis of some kind is not necessary. However, if the bush is not recognizable, research is required. This might be referring to a book, or perhaps a recognized expert. Next time, you have this newly acquired pattern stored in memory and immediate recognition occurs.

An important question concerns the personal qualities that are necessary to become, in this case, an expert farm manager. While training and experience can make up for a lack of inherent ability, a good measure of both is probably highly beneficial. Being an expert enables appropriate decisions in good time. Good farmers seldom have to spend large chunks of time to sort out a problem for their systems have the right answer stored. This ability to quickly make a decision is sometimes referred to as 'intuition', or sometimes 'tacit knowledge'. This sounds like a mysterious quality that only some managers have. In fact, it probably relates to having the right attributes, experience and training. Thus, intuition is probably a learnt attribute that we all have to a greater, or lesser, extent. As it is not appropriate in many cases to spend a long time researching a problem or opportunity, developing this intuition is an important aspect of being a good manager. Similarly, knowing when your intuition is likely to be incorrect is also obviously an important attribute. In such cases, formal study, research and analysis are required. This usually leads to an enhancement of the manager's intuition.

Skill and intuition must cover a wide range of areas for successful farm management. Of any production systems, agriculture, and horticulture, involves an extremely wide range of necessary skills. Frequently production involves:

- soils, rainfall and climate, plants;
- animals, harvesting and machinery in general, engineering (buildings, structures, irrigation...);
- labour and personalities;
- markets, finance and economics;
- politics and the resulting impacts on the rules and regulations that must be complied with.

This very wide range of subjects covers everything from physics and chemistry through to biology and psychology in that the people involved operate within the bounds of their human characteristics. The excellent manager will have a reasonable understanding of all these areas.

Nevertheless, equally, if not more important, are the management skills that a manager brings to the job, which, in turn, lead to the decisions made, and implemented. Thus, a manager must know how to use facts and figures through being skilled in:

- *understanding* the technology and what lies behind it (sowing rates, fertilizer outcomes, the sciences – biology, physics, etc. involved);
- *observation and recording* (soil conditions through to international markets);
- *planning* (risk management, cash flows, job priorities, time management, economic principles, etc.);
- *anticipation* (possible outcomes and their chances);
- people skills (labour management, network maintenance, negotiations, etc.);
- personality management (stress management, motivation, objectives, and so on).

Acquiring abilities in all these areas is dependent on the basic attributes of a potential manager, and the opportunities for training and experience that are made use of. Some will be good at a sub-sample of the attributes, and some will be rounded with a complete package of excellent skills. The critical question relates to how a manager might acquire such a fully rounded set of attributes.

It appears that a human being is defined by two, possibly three, basic sets of factors:

- the first is a person's personality;
- the second their intelligence; and
- the third their motivation, although some researchers believe motivation arises from their personality and intelligence.

Personality is made up of sub-factors such as extroversion and anxiety levels, and intelligence is similarly made up of components such as memory and reasoning. In each segment, an individual will have a defined make-up, leading to the unique whole. Some will be good managers, others not.

When observing the attributes of a manager, you observe what is called his or her 'phenotype'. In contrast, the package that a person is born with is referred to as the 'genotype'. This is defined by the inherited genes passed on 50–50 by the parents. But the genes only define the person's building blocks, which then interact with the environment and experience that a person is exposed to. The sum of the genotype and environmental experiences gives rise to the observable phenotype.

The question is what phenotype is appropriate for good management, and how is this attained? Clearly the genotype cannot be altered, nor can a person's early experiences – this is now history. However, some phenotypes will more than likely be capable of being modified through the correct training and experiential exposure. For a manager to improve his or her skills, it is clearly important to discover what training will in fact work. This assumes phenotypes are alterable, and fortunately there is evidence this is indeed the case. Understanding phenotypes and managerial skill is also important in setting up the correct conditions right from an early age to ensure that good managers are created. Clearly, the approach of providing an appropriate set of experiences and training in early life is preferable to trying to improve skills in later life, though both possibilities must be followed.

People, and managers, are seldom totally consistent in their actions through time. This might be because they have re-evaluated their objectives, but more likely simply because they are human. While an individual has fixed traits defining his personality and intelligence, how these give rise to decisions made on any one day can be variable. You will be aware that your emotions and state of mind vary from time to time in response to events and activities. This is an expression of personality with some people varying more than others as defined by their phenotype. The consequence is that the decisions made will not always be consistent, even given the same circumstances, particularly for people with an anxious personality. Effectively, a person's state varies, so the astute observer must be aware of this and not conclude on a person's managerial ability based on a small number of observations.

While learning the basics of what gives rise to a person's actions is valuable in itself, the important aspect of this understanding is considering how managerial skill might be improved. Most managers improve with time as they take in the lessons of experience, but the extent of improvement probably depends on their personality and intelligence. This situation has been expressed in Fig. 1.1.

At any given point in time, a manager embodies a particular skill level that is dependant on the past environment, intelligence, age and personality. Also relevant are the manager's objectives in that, for example, he might not consider efficiency as being relevant. Also possibly relevant is what is known as a manger's 'locus of control'. This reflects the attitude to how much control over the farm's destiny the manager believes he has. Some believe, for example, that outcomes are largely determined by the weather and markets and, therefore, managerial skill is not that important.

Then, through time, change in skill is dependent on the training and its effectiveness. Success may well depend on the type of approach the farmer prefers relative to the structure of any training offered (preferred learning style). Precursors to formal training, if any, are the manager's desire to involve himself in training, and then his potential to actually absorb and learn from the training. Often, the training is totally informal and relies on the farmer's reading, observation and ability to learn from mistakes and experiences. Understanding the personality and intelligence of a farmer helps understand whether a farmer will improve, and the best way of undertaking this improvement.

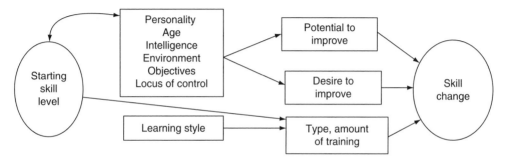

Fig. 1.1. A transactional model of improving managerial skill.

Of course any formal training comes at a cost. There might be formal tuition fees, and certainly there is a time commitment with the attendant opportunity cost. There may well be an emotional cost as well in that 'going back to school' can be traumatic. Some farmers will not be prepared to accept the discipline involved in attending evening classes after a day of physical activity, and commit time to working through exercises. Besides the emotional and discipline issues, the decision to take up formal improvement also involves economic issues. The decision is very much a matter of assessing the costs relative to the gains from the improved management. However, it does not take much improvement to cover the time cost, particularly where the training improves the farmer's intuition that will have ongoing value up to retirement.

Historically, farm management researchers and writers have commented on the importance of managerial skill, but this aspect of production efficiency is seldom highlighted, and the level of research funds devoted to the area is quite minimal. This situation needs to change as any manager is clearly the key to combining resources appropriately to achieve the farm goals.

Early texts, for example Case and Johnson (1953), note the importance of the manager, and from time to time research on the management process has been published. Research examples include Johnson *et al.* (1961), who studied Midwestern farmers in the USA with respect to how they operated; Ohlmer *et al.* (1998) similarly studied Swedish farmers and concluded that their decision process was dynamic with constant adjustments and reviews.

Furthermore, there have been a range of studies over the years that have related human factors with success and efficiency. The factors used in the analyses have ranged from the farmer's age and education through to psychological measures. Early work considering the factors was reviewed by Muggen (1969), who concluded that there were 61 variables correlated with success. Figure 1.2 gives his summary of the situation. The variables included socioeconomic status, education, motivation, vocabulary, agricultural knowledge and many more.

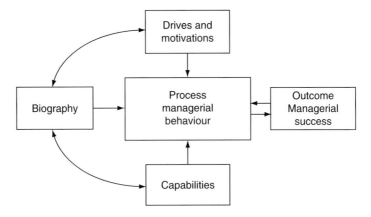

Fig. 1.2. Human factors in farm management. (From Muggen, 1969.)

Other studies looked at:

- human capital relative to farm size and growth (Summer and Lieby, 1987);
- psychological type and financial success (Jose and Crumly, 1993);
- thinking styles and financial characteristics (Howard *et al.*, 1997);
- experience and production efficiency (Wilson *et al.*, 1998);
- personality and intelligence relative, in particular, to environmental behaviour (Austin *et al.*, 1998);
- defining a farmer's management capacity (Rougour *et al.*, 1998);
- measuring managerial efficiency using personological variables (Warren *et al.*, 1974; Tripp *et al.*, 2000);
- personality measures, managerial ability and dairy production (Young and Walters, 2002), and
- predicting farm management performance using farmer decision-making profiles (Solano *et al.*, 2006).

These studies show that management skill is indeed correlated with many factors that together define a farmer's managerial ability, or what can be called 'human capital'. The limited amount of research shows there is still a general lack of recognition of the importance of studying and understanding managerial ability. And, furthermore, the studies have not provided a basic theory and understanding of the creation of a farmer's skill, or lack of it, nor of how an individual's skill level can be improved.

The following chapters bring together this, and new, work as well as developments and theories from other disciplines, including psychology, to provide a well-structured set of ideas for understanding managerial ability. Chapter 2 contains outlines of the characteristics of a farmer that defines his modus operandi, and how these characteristics might be measured using psychometric tests. This is a person-centred approach in contrast to the use of production parameters that are the outcomes of the decisions made.

Chapter 3 looks at the relationships between the factors that define a manager. What is the influence of each in creating the success, or otherwise, of the managerial ability? Clearly knowing the importance of the precursors to ability enables recognizing the areas that might be improved with reward. Effectively, the discussion covers the 'origins of ability'.

Also important is an understanding of the processes involved in comparing alternative decisions, and therefore, actually making decisions. Consequently, Chapter 4 is devoted to reviewing the possible decision criteria and processes used. These criteria range from the concept of maximizing a general measure of output, utility, through to the simple approach of producing a minimum requirement of an output such as leisure time based on intuitive feelings.

Having covered the factors giving rise to ability, Chapter 5 contains an outline of the competencies necessary for successful farm management. Obvious areas are skills such as accurate observation, and an ability to forecast or anticipate outcomes following specified decisions. These and other competencies are reviewed using farmers' views.

In that farmers are seldom rational machines in their decision making, there is an implication that their decisions can often be biased, that is, incorrect or wrong relative to their objectives. Thus, Chapter 6 outlines the common errors, or biases, and, therefore, highlights areas that require special attention when improving ability.

Most farmers worldwide are owner–operators in various forms. The essence of this is that a farm family is intimately involved in farm life and decisions with the farmer and his or her family living and working on the farm. Consequently any study of a farmer must also involve a study of the family and their influence on the objectives and decisions taken. Chapter 7 considers these influences and also the factors that give rise to the objectives held by the farmer–household complex for success in farming must revolve around having both the objective set that is right for the family, and having systems that enhance the attainment of these objectives.

Chapter 8 then reviews whether success is possible from using improvement programmes. If, in fact, the precursors of ability are fixed for all time following birth and early experiences, there would be no point in training schedules. This is not, however, the case.

Finally, the various threads of the arguments are brought together to provide conclusions.

In summary, this book provides an understanding of a manager's characteristics and why a particular person reacts in a defined way when faced with a range of situations. The important features relate to the nuances and attributes of a person's expressed personality and intelligence together with how the person's persona impacts on such things as their learning style, motivation, objectives and belief in their control of their farm management destinies. As the majority of farm situations involve an owner–manager arrangement, the farm family is also an influence on the outcomes and their efficiency.

Besides using casual observation to assess a manager's characteristics, this book also provides written tests to gauge personality, intelligence and the locus of control. Benchmark data for a sample of farmers is also provided. It also provides the results of surveys to assess what farmers consider as the important skills associated with successful management. Question sets are also provided to assess a person's objectives, the important physical and managerial aspects of a farm for there is an intimate association between a manager, and his family, and the farm itself (as in Fig. 1.3) which can influence the skills required.

Comments are given on how to improve ability following a discussion on whether improvement in adults is possible. This relates to the argument on 'plaster or plasticity'. Is management stuck in plaster, or can it be changed?

The contents of the book do not cover all skills that might be important, as there are many texts available on things like negotiation skills, labour management, listening skills, production economics and related computer packages covering, for example, mathematical programming and optimal seeking systems, regression analysis, cost–benefit analysis, and the like.

However, what have been included are the comments and thoughts of two farmers whose situations have been used to demonstrate the various fac-

Fig. 1.3. Being a farmer confers many benefits. The beauty and remarkableness of the farm confers constant wonderment.

tors raised in the book. In addition, for the chapter on the methods of improving managerial skills, the experience of two farm consultants has been explored to obtain their views. Details of these two consultants are provided in the chapter on the skills required as consultants also have their views on the important attributes. But it is appropriate to introduce the two case farmers at this point.

The Case Study Farmers

Introducing Margrave

The first farmer has suggested he be known as Margrave. Perhaps his ancestors had connections to the Margraves of olden times for there are several castles and other sites of this name in Europe and Britain. This only emphasizes that Margrave has had connections to farming for centuries through his ancestors that he has been able to trace back to 1066, with every generation being a farmer in some form. Therefore, you have to say farming is in his blood. Margrave manages what even in western country terms would be called a large operation of some 3000 ha with eight full-time employees, though 300 ha are devoted to forestry that requires less than constant attention with activities such as pruning being able to be conducted at times that fit in with other farm activities. However, he was not always involved in such a large investment for he first started farming on 180 ha before progressing to another farm, and finally ending up on his present operation. Margrave's farm is a mixture of river flats and hill country with several neighbouring vineyards and cash cropping areas on the river flats, so some of the area is of good

soil fertility, but the hills are less productive. Margrave's flat land is used extensively for producing animal feed crops for winter use in the temperate climate
that enables the stock to remain outdoors all year. The annual rainfall is
approximately 980 mm but with quite wide variations from year to year.
Indeed, one year a cyclone produced huge volumes of rain and much flooding
and subsequent damage, but of more frequency are the lower rainfall sum-
mers. Normally, summer rainfall is low, but not infrequently the rain just fails
completely, and so droughts can be a major problem with totally inadequate
pasture production for the animals on hand. This causes many decision problems – what do you do? Sell off some animals today, or will the rains come
tomorrow, saving the day?

Total stock involves some 33,000 stock units, where a unit is defined as
the pasture needed to support a 50 kg breeding sheep (ewe), but this is made
up of a mixture of animal types. There are 10,000 ewes and 3300 hoggets
(young female sheep), with the ewes producing 125 lambs per 100 ewes,
(Fig. 1.4) and the mated hoggets producing 75% of lambs. In addition, there
are 1000 beef cows (Aberdeen Angus) with replacement young stock. Indeed,
all replacement young stock are bred on the farm, but in addition some 10,000
extra lambs are purchased each summer (which are fed up to a saleable weight
for meat) as well as 300 yearling (12 months old) bulls which are similarly fed
up for sale as meat animals. Some of the breeding cows are mated for calving
in the autumn to catch a high price market, but the rest calve in the spring. The
large numbers of animals purchased in the summer enable greater flexibility in
the system, so if the summer has particularly bad rainfall, the numbers pur-

Fig. 1.4. Each season brings new hope. A scene typical of what inspires Margrave
in his fascination with agriculture.

Fig. 1.5. Large sheep flocks can be somewhat daunting, but not for Margrave. A flock being brought in for attention.

chased can be reduced. However, in normal times, if there is such a thing as a normal season, all progeny and purchased animals are fed to a weight suitable for meat production.

Margrave has been a manager for 14 years (Fig. 1.5), following 5 years working on farms. Most of his life has been rurally orientated, with living on a farm from birth and attending a rural primary school. However, he was sent to an urban boarding school for his secondary education, which he notes 'was an experience, but I didn't work much relying on natural ability to get me through'. While his life centred on the country, Margrave was not really exposed to the decision-making side of agriculture with his family keeping discussions about farming and family matters to themselves, something that was not uncommon in earlier times, 'private matters were not discussed with me.' While Margrave did help with farm jobs, he notes 'it wasn't until I was 16 that I learnt much about agriculture'. He left school about this time.

Introducing Hank

The second case study is also about a stock farmer, but operates a totally different system, and also has a totally different background. He operates a dairy farm, well in fact more than one, and while his early experiences were on a family farm, his initial working life was spent in the banking world. This very successful dairy farmer wants to be called Hank despite not being originally from The Netherlands. In fact, he was born and bred on the west coast of the USA. Margrave, on the other hand, was born and bred in New Zealand. Hank talks about his ancestors and noted that they too were involved in farming in the Midwest of the USA. One grandmother was so attached to the land that when she died she wanted her body to be shipped back to the Midwest for burial. This strong attachment is not at all uncommon, as a farmer cares for his land with a passion for many years and sees it change as the decades pass. His heart and soul is often tied up in the life-giving material upon which we all gaze from time to time.

Hank and his wife, who we will call Hanna, started farming on their own account in 1987, and ever since then they have grown their investment by acquiring more land and indeed only recently purchased an additional 145 ha for conversion to a dairy farm. As noted, Hank was born in the USA, whereas Hanna is a New Zealander. Hank's father was both a dairy farmer and golf course owner, so he was probably an entrepreneur. Hank worked on the farm as a child and has very fond memories of his father, with many long chats about farming and decision making. They also had special times together travelling to other farms to inspect their systems. They were, clearly, the great-est of buddies. However, Hank's father died when he was just 16, and his much older sisters became important in his life. Yet, the seeds of a passion for dairy farming had been well and truly sown. Hank eventually ended up at a major land grant university, studying agriculture that led him into the banking world where he spent his time assessing loan requests from farmers and over-seeing their use of the money. He enjoyed this job, and quickly rose to become a vice-president at the age of 33 years. Part of his education involved a visit to New Zealand on a university-exchange programme. He is not sure why he applied to be one of the first on the programme, but he ended up being selected and travelled to Lincoln University, which is where he met his wife-to-be. Later, Hanna travelled to Hank's American university on the reverse exchange programme.

Hank maintained contacts in New Zealand, and even bought calves to be contract-reared from a distance. One day a friend in New Zealand suggested that he apply for a share-milking (manager's) position, and buy cows for the prices had reached rock bottom with major changes in government policy. As Hank could not see himself working the rest of his days in a bank, and having observed how his older colleagues were locked into the system with pension plans that meant they could not leave without major financial sacrifice, he applied for the job. Fortunately Hank was successful (he suggests the state of agriculture at the time meant he was the only applicant!). This start in farming was in part possible due to the cows Hank had in New Zealand that were reared by 'remote control'. Acquiring cows, and eventually land, together with being a manager, led to Hank and Hanna setting up their own farm and the significant holdings they now managed (Fig. 1.6).

Hanna and Hank now operate two dairy farms using various ownership structures. One consists of 200 ha fully irrigated, milking nearly 700 cows. The other is just starting up, and consists of 136 effective ha. This block has 500 cows. In addition, 80 ha of flats bordering a major snow-fed river are leased and used for feed production. Twenty hectares of this block are irrigated. The total

Fig. 1.6. Modern dairy farming is capital intensive. A scene typical of Hanna and Hank's investment.

production system is run with eight labour units. The milking and day-to-day management on the main farm is carried out by a share milker, who shares both the costs and profits, and the other block has a young manager who has some, but lesser, profit-share arrangements. Both their shares vary a little depending on the price paid by the milk company for the milk solids produced. If there is a downturn, everyone helps take the brunt, and vice versa, though there is a baseline system to ensure the share milkers have at least a minimum income. In recent years, the milk company payouts have been reasonable. With the development of improved pasture and systems, including the use of new irrigation equipment, production has steadily improved. Ten years ago production per cow was 383 kg of milk solids, and 1044 kg/ha. Now, the per-cow figure is more like 420, but more importantly, the per-hectare production has increased to 1420 kg. While the soils are relatively thin, they at least have good drainage, and with the constant application of fertilizer and improved plant cultivars, the organic matter has been building up leading to higher and more robust feed production. In the temperate climate, the cows are outdoors all year (Fig. 1.7), and usually milk 10 months of the year. Hank was particularly keen on getting away from all the expensive and energy-hungry housing systems necessary in many areas of the USA. Over the years, the genetics of the animals has improved with careful selection of the sires through the artificial mating system. Replacement heifers were raised by Hanna and Hank until a share milker was employed, at which time he took over the rearing. However, with the purchase of the new farm, which was originally running sheep, the breeding programme has started up again.

Fig. 1.7. Skill levels must be high to obtain a return from intensive dairy production. Part of a typical herd belonging to dairy farmers such as Hank and Hanna.

Hanna and Hank started with very little, and are proof that careful, but also excellent, management can lead to a major investment with high economic production. Their human story is worth listening to.

As appropriate, the stories of the two case farmers will be introduced and discussed in each chapter with respect to the particular issues raised. They provide both concrete and human evidence of the issues presented.

2   What Defines Management Ability?

Introduction

Anyone can be a manager – but they may not be particularly good at achieving their objectives. The desire to be a manager is relevant in success. Some farmers acquire their status due to tradition and the handing over of assets, rather than a keen desire to make a career from managing primary production. On the other hand, some who want to be farm owners and managers find it is impossible due to the resources required. Whatever the case, an ability to take and accept the responsibility of making and carrying out decisions is an important precursor, but whether a person can in fact make good decisions relevant to a particular farm (for each is unique) depends on whether they have the required attributes and experience. Abilities such as making sure that the jobs are carried out in a timely manner (e.g. spraying weeds before they are too mature; getting supplies delivered before it is too late to complete the job; marketing the product before the prices drop, and so on) are crucial and probably relate to a person's degree of conscientiousness as well as an understanding of the biology involved. Conscientiousness is a personality trait. Similar examples exist for the other attributes so a manager's personality impacts on their likely success as a manager.

Management also involves holding in memory a large store of recallable information – thus a good memory is a valuable attribute. It is also important to be able to sort the grain from the chaff when considering, for example, inputs being offered by a myriad of sales people. Fertilizer choice is a prime example. Getting it right requires clear and logical thinking, using knowledge of soils, fertilizers and nutrient requirements. These two examples relate to a manager's level of intelligence, so, as you would expect, reasonable intelligence is a requirement for good management. This does not necessarily mean high achievement in formal education as there are many examples of excellent managers who have had little schooling. On the other hand, there are an increasing number of managers in the developed world holding tertiary qualifications.

Even given the right personality and good intelligence, a new manager will struggle to get it right if lessons have not been learnt from experience. Farming is a very practical occupation where a good knowledge of practice is important. A good understanding of systems and outcomes is most unlikely without considerable experience and lessons from past mistakes. Being involved in situations that cover the problems and conditions likely on a farm is important, as is an ability to self-analyse and benefit from the experiences. Learning from experience is probably dependent on good intelligence and observation systems. It should also be noted that while some people can not formally analyse what they have learnt, their intuition may be finely tuned having inherently picked up the lessons on offer.

A manager must be very clear over what his responsibilities are. Figure 2.1 outlines these in a general sense. Action and decisions always starts with planning, the conclusion from which might be to actually take no action. Thus, a review, for example, of whether to harvest a crop might conclude that it is not yet ready. The plan is 'wait and review'. Other plans where action is required then move on to the execution phase. The plan must be put into effect. This involves, potentially, the myriad of actions necessary in getting a plan completed. Planting a crop, say, requires ordering and receiving all the seed, fertilizer, fuel and other supplies; checking the operational status of the equipment required; getting the labour organized; and, finally, actually getting the seed into the ground at the right depth. Prior to this, of course, the area had to be appropriately prepared (through cultivation perhaps).

Once the plans have been activated, the next phase involves 'control'. The outcomes must be observed to ensure that the expectations occur, and where a variation is observed, replanning may be appropriate, and so the cycle repeats itself. For example, as the seed is germinating, a vigorous batch of weeds is also striking that require spraying or, perhaps, inter-row cultivation.

This whole planning–execution–control cycle is a very dynamic operation, with constant observation and replanning leading to further execution. Particularly in agriculture, plans are seldom completed as originally expected due to the constant vagaries of the weather, markets and other risk factors such as the availability of suitable labour. A successful farmer must recognize this dynamic process,

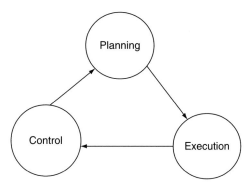

Fig. 2.1. Responsibilities of management.

and always be prepared to reanalyse and change plans and actions. This is what makes it a wonderful challenge in which only the skilled really succeed. It is easy to see why an appropriate personality and good intelligence is highly desirable. However, any manager can have a particular phenotype that may not be what is best for good management, so in these cases it is important to recognize this and make efforts to make as many modifications as possible.

This improvement process is helped if farmers are able to classify themselves, or seek assistance from a professional to determine where they might change to improve their management. Absorbing such changes to provide permanent improvement is desirable, given that most of the planning–execution–control process is carried out intuitively, in contrast to a formal paper-based operation. Whether, and when, to drench a mob of cattle for internal parasites, for example, will be an intuitive decision based on careful observation of the animals, which may include faecal egg counts, and noting the feed supplies and likely absence of larvae and eggs on alternative grazing areas.

Classification of farmers into skill groups can also be intuitive, or more formally based on observing outcomes. The ultimate criterion is the 'end-of-year profit' relative to peers on similar farms. Alternatively, it is possible to administer written tests, the results of which can be assessed against benchmarks. The reasons for tests include providing a knowledge of the farmer's current ability, and thus provide a baseline from which changes can be observed. This baseline test also helps indicate whether improvement is a real possibility. If the farmer is already exhibiting all the right characteristics, then only fine-tuning will be possible, perhaps in conjunction with a professional consultant. The tests can also suggest the particular areas that might be worked on for improvement. For example, a specific farmer might exhibit a high level of anxiety, resulting in apparent stress and inappropriate decisions that, perhaps, are designed to remove the chance of risky outcomes. If this proves to be the case, a programme designed to dampen the anxiety could be appropriate. Similarly, it is useful to know a farmer's strengths so that they can be exploited. For example, a strongly extroverted person may well be able to excellently manage a committed work team, enabling, say, intensive cropping using a large workforce for good reward.

Tests such as those discussed are generally called psychometric tests in that they are designed to explore a person's psychology, that is, to find out the characteristics that determine how a person will react under a range of circumstances. The study of psychology is the study of people and how they react to their environment. This is exactly what we are talking about – how managers react following observed cues from around their farms, markets and the wider community. Psychologists have developed a large number of tests to explore a wide range of personal characteristics. Most people have experienced an intelligence test at some stage in their life. These tests are mainly designed for the general public, whereas in agriculture there are specific attributes that are important. The development of suitable tests for assessing primary producers is only in its infancy, with limited experience available. Later in this chapter some of the tests that have been developed are provided together with benchmark data. In a business sense, psychometric tests for non-rural businesses have been used for many years, but most are not suitable for agriculture, as the situations are quite

different. These tests continue to be used in urban situations, indicating they
have been valuable. Many people will have experienced at least one of these
when thinking about the occupation they would like to spend their life in.

As noted, the two critical aspects of a person's management are their personality and intelligence. The next sections in this chapter cover the important
aspects of these two broad features of a person, and introduce tests that can be
used to assess them. The 'personality' test is actually slightly different from a
standard test in that it is called a 'management-style' test, as this is the aspect
of personality that we are interested in. While modern personality theory suggests that people have five basic traits, the management-style test looks at six
factors. Similarly, the 'intelligence' test outlined is referred to as an 'aptitude'
test, for it is a person's managerial aptitude that is important. Again, the components of this test relate to the intelligence aspects that are likely to be important in farm management.

Other tests that can be important include the 'locus of control', reflecting a
farmer's belief in how much control he believes he has over outcomes. Assuming
control is possible, a belief that you can influence outcomes is obviously important. A test for the 'locus of control' is provided in Chapter 3. Also relevant is a
farmer's preferred way to learn new skills in that some people do have abstract-
ing abilities and can learn from books in contrast to needing practical demonstrations and experience. A test for learning style is available and will be
discussed.

As farming is a very risky business it is important for a farmer to be fully
aware of their attitude to risky decisions. Thus, tests to determine risk attitude
are introduced. With intensive effort it is possible to alter these attitudes with
some benefit. A farmer's objectives also relate to this question, therefore a brief
discussion on objectives is also included, which is expanded in a later chapter.
Similarly, motivation can be an important factor related to objectives, so some
comments are provided on this subject. Finally, a test called 'emotional intelligence' is introduced because it is important in urban business, but its application in agriculture is probably much less significant.

Personality and Its Links to Management

While many ideas on personality have been suggested over the years, and this
debate still continues, many psychological researchers consider the five-trait
model as being correct. One text that describes the model is provided by
Matthews and Deary (1998) with references to many of the basic research
articles giving rise to the five-factor model. While each base factor can be considered made up of many facets, the core traits are:

- 'openness';
- 'conscientiousness';
- 'extroversion';
- 'agreeableness'; and
- 'neuroticism'.

It will be noted the first letter of each makes up the word OCEAN. While the word neuroticism is commonly used, in a management sense probably a better word is 'anxiety'. Neuroticism has connotations of the extreme ends of the spectrum. Although we all have a particular level of this trait, it is likely to be far from the extremes and, therefore, make us anxious rather than neurotic in the everyday use of the word.

The openness trait is expressed through a person's attitude to new ideas. A particularly open person will be daring, liberal and somewhat original in their thinking. In contrast, a person who scores poorly on the openness scale will be conservative, unadventurous and conventional. It is clear that it is desirable for farmers to have the right degree of openness to suit the situation they operate under. In an environment where change and opportunity are available, a high degree of openness will enable capturing the potential benefits of new products, new methods and systems. On the other hand, where the markets, technology and political conditions are relatively constant, lack of openness is not much of a problem. On balance, provided an 'open' person does not get bored, a farmer who is very 'open' is more likely to succeed, given that new products, ideas and opportunities are always likely at some stage in a farmer's history of work. In real life, farmers will exhibit a full spectrum of degrees of openness. A farmer's suitableness for the each environmental situation will similarly follow a full range.

Conscientiousness is clearly an important trait for successful management. Someone who has a high rating on this trait will be careful, reliable and takes responsibility seriously. Such people can generally be relied upon, and when a task is agreed you can be sure it will get done. In contrast, someone who exhibits the other end of this trait's scale will be somewhat careless, undependable and even negligent. While, no doubt, there are some successful managers who show little of this trait, possibly through good luck, generally a good level of conscientiousness will be associated with successful managers. On the other hand, too much conscientiousness could be counterproductive, particularly when associated with a low level of openness. Being extremely careful, with little daring, could well stifle good management. This idea starts to indicate just how complex we are. Once two traits are considered, each of which can have a wide range of levels, a specific combination will have a specific relevance to managerial skill. Once all five traits are considered, it is highly likely that any one farmer will be unique relative to all other farmers. Tests, as discussed below, enable sorting out where a farmer lies in the full spectrum of possibilities.

Extroversion is a trait which most people recognize quickly. An extrovert is sociable, talkative and frequently spontaneous. You probably know people like these, and perhaps you are an extrovert as shown by how you seem to be energized when talking to a group of people. Extroverts will enjoy both farmer and social events in contrast to the opposite personality, referred to as an 'introvert'. Such people are probably shy and find mixing with many people difficult and not to be courted. Thus, an introvert will be retiring, quiet and somewhat inhibited. As a factor in good managerial skill, a degree of extroversion is necessary in that it is important to interact easily with people when looking for information and help. An introvert will be backward in seeking help, be it in a professional sense or in a personal help situation. Also important is how well a person manages staff and contractors. An extrovert will generally find it easier to express requirements and feelings, both being important aspects in running a successful team. In a nutshell, extroversion is probably a factor in leadership skills, though other attributes are no doubt also relevant. On the other hand, for farms where a big team is not required, a degree of introversion is probably quite acceptable for good management.

Agreeableness is the fourth basic trait. A 'basic trait' is a distinct feature that is independent of all other traits. The research that suggests these five basic traits considered correlations between a very wide range of personality features, and discovered that many of them were correlated. If the correlated groups are separated, the OCEAN traits are left as independent. A person classified as being 'agreeable' is good-natured, soft-hearted and somewhat selfless. Generally these people might be called benign and seldom get angry or overly exited about issues. At the other end of the spectrum, a 'non-agreeable' person will be rather irritable, and certainly ruthless as well as being somewhat selfish. It will be noted that these characteristics are somewhat disparate, but research has indeed found that they tend to be highly correlated. In a management sense, a certain degree of 'agreeableness' is probably helpful in that a calm and accepting nature helps handle the ups and downs of working in an environment over which there is little control. When the hail storm wrecks the crop just before harvest it takes a very agreeable person to maintain an objective view of the situation. Similarly, when the farmhand reports he has driven the tractor into a fence post destroying the fence and denting the equipment, an agreeable outlook is most helpful.

On the other hand, there are situations where it does not pay to be agreeable. Sometimes getting a little irate with a dealer, or perhaps a contractor, when the job is not being done as required can achieve some much-needed action at the right time. Thus, a personality that is on the agreeable side of the middle line is helpful, provided assertiveness can be produced when some positive action is clearly necessary.

Finally, there is the 'anxiety' trait (or neuroticism as defined by psychologists). An anxious person will worry, be somewhat vulnerable to less than favourable outcomes in that they will easily become emotionally upset and also exhibit insecurity. That is, they will tend to be unsure of whether they made the correct decision and often worry endlessly over the decision after the event even though they know full well the die is cast and nothing further can be done about the situation. Most people will have at least one acquaintance that can be called anxious. In the extreme, anxiety can lead to physical illness and other undesirable characteristics like endlessly going back to check, for example, whether the stable door has been successfully locked.

A non-anxious person will, on the other hand, tend to be calm, resilient in contrast to vulnerable and feel relatively secure in the decisions made. In that anxiety can lead to emotional upsets it may well give rise to irrational decisions and distractions. A farmer at the anxious end of the spectrum is less likely to be an excellent manager, and certainly not one that is at ease and enjoying the work. It might also be thought that anxiety and conscientiousness might be related, but they are in fact two separate and independent traits so it is possible to have an un-conscientious person, who is very anxious in that they might fail to carry out a job on time, and then spend the rest of the week worrying about the consequences.

Anxious people are also likely to avoid risky situations, though this approach is also related to the openness trait. Most farmers will tend to be what are called 'risk averters' in contrast to 'risk preferers'. Risk averters will choose options, where there is a choice, that tend to have a more clearly definable and known outcome. For example, if a fixed price contract is available for pre-selling a crop yet to be harvested, in contrast to waiting to see what the open market brings, they will tend to take out a contract. This fixes the price they will receive. Of course, the price offered relative to the estimated open market price will also be a factor with a decision, depending also on the degree of risk aversion exhibited by the farmer. On the other hand, a risk preferer may well choose the open market situation, though again this will depend on the relative price forecast and the degree of risk preference. It is relatively easy to find out a farmer's risk attitude through tests which are referred to below.

While the five-factor model of personality is generally accepted as the most useful in modern psychology, other models do exist. For example the Myers–Briggs model (Keirsey, 1998) is popular and readily available. This model assumes people can be divided into 16 types based on the factors:

* extroverted (E)–introverted (I);
* sensory (S)–intuitive (N);
* thinking (T)–feeling (F); and
* judging (J)–perceiving (P).

Each pair expresses the extremes in the factors, so people are classified according to which end of each they fall. The four types then give rise to 16 possible combinations, so a person might be, for example, ENTJ in which case they are extroverted, intuitive, thinking and judging. A simple test enables classification.

Examples of the questions in the Myers–Briggs test include: 'Does interacting with strangers energize you, OR tax your reserves?', 'Are you more comfortable after a decision, OR before a decision?' and 'Common sense is usually reliable OR frequently questionable?'. In total there are 70 of these questions. The answers tell which of the extremes of each of the four factors the person is. A template on which the answers are recorded is used to add the scores and provide a type classification. Some of the earlier work relating personality to farm management used this test (e.g. Young and Walters, 2002).

Still, other models have been proposed. An example is the frequently used 16-factor personality factor (PF) model, which does have strong correlations with the five-factor model (Cattell et al., 1970). Examples of the traits are:

* outgoing–reserved;
* unemotional–emotional;
* cheerful–sober;
* tense–relaxed;

- conscientious–expedient;
- radical–conservative; and so on.

The similarity with the five-factor model is clear. Contemporary research suggests that many of the 16 factors are highly correlated with one of the five factors so that in a search for parsimony and the core traits, the conclusion has been that the five-factor model is more robust and useful.

Personality and Tests

Determining a person's personality can be approached in two ways. Careful observation can give many clues, especially if the observation can continue over a year or so during a professional relationship, and the observer knows what to look for. Anxiety, for example, tends to become obvious through discussions over the selection of alternatives, and through general discussion over outcomes. Similar comments apply for assessing a farmer's risk attitude and the other personality traits.

The other alternative that enables assessing the degree of the various traits is to use a formal question-and-answer test, though such pencil-and-paper tests can be misleading when they are not worded well and the farmer finds it difficult to express his real feelings and attitudes. In the end, use of a formal test as well as astute observation will provide the best of both worlds. Formal tests do not provide absolute measures, for there is no absolute benchmark like there is for things such as distance (the metre standard held in Paris, for example). However, data from a wide range of farmers provide relative benchmarks.

As noted, there are many personality tests based on the various theories of personality. For the five-factor model, the common test is referred to as the NEO-Personality Inventory – Revised (NEO-PI-R) (Costa and McCrae, 1992). This test has 240 questions, 48 for each of the five factors. To administer the test you must normally be a registered psychologist before the material will be licensed to you. Given the length of the test, and the copyrighting associated with it, including the payment required for licensing, as well as the fact that it is designed for general use in contrast to being specific to farmers and the terms they specifically relate to, it is not very practical for farmer use. Accordingly, a shorter and farmer-directed test has been developed and tested. This is described below. It must be noted that there is a shorter version of the NEO called the NEO Five-Factor Inventory (NEO-FFI) made up of 60 questions with 12 for each factor.

With a shorter test, some of the nuances of personality are lost, but when considering farmer managerial ability, these nuances are not so important. The five-factor model has what are called 'facets' for each base trait which the long tests measure. The facets include:

- Openness: fantasy, aesthetics, feelings, actions, ideas, values.
- Conscientiousness: competence, order, dutifulness, achievement striving, self-discipline, deliberation.

- Extroversion: warmth, gregariousness, assertiveness, activity, excitement seeking, positive emotions.
- Agreeableness: trust, straightforwardness, altruism, compliance, modesty, tender-mindedness.
- Neuroticism: anxiety, angry hostility, depression, self-consciousness, impulsiveness, vulnerability.

It will be noted that some of the facets are unlikely to be strongly related to managerial ability. Examples include modesty, aesthetics, altruism, tender-mindedness and excitement seeking.

It was considered that a farmer would have little difficulty in answering 25 questions (actually statements) given that in any test situation there would also be other tests included. Thus, a test of five questions for each basic trait was constructed and given to several samples of farmers for testing.

Given that the objective was to assess a farmer's personality with respect to management, the test was referred to as a management style test. This contains 25 statements each of which the farmer is asked to rate for the degree of truth on a 1 to 5 scale (true = 1, … untrue = 5). The questions are presented in Appendix 2A to this chapter.

Given that the data was collected from many hundreds of farmers, the correlations between the questions were assessed to determine the underlying factors. The results clearly indicated there were six management-orientated traits that reflected a farmer's management style (Nuthall, 2006).

Based on the questions making up each trait, they were called:

Questions making up the trait (no. in question set)		Average score
Correctness anxiety	(5, 10, 13, 20)	12.8
Conscientious planning	(8, 9, 16, 17, 19, 21, 24, 25)	20.4
Thoughtful creativity	(2, 9, 11, 12, 15, 16)	14.3
Community spirit	(7, 15, 18, 23)	11.2
Consultative logician	(1, 2, 3, 4)	9.9
Benign acceptance	(6, 7, 14, 22)	11.8

Beside each trait is listed the questions that make up the factor and the average score for the sample of farmers across all farm types. The score is obtained by adding the statement 'truth score' (1 to 5). The averages vary due to both the tendency to be at one end of the scale and the number of questions making up each trait. It will be noted that farmers, on average, do not tend towards exhibiting 'correctness anxiety'(maximum possible score is 20 for completely 'not true'), are middle of the road for 'conscientious planning', definitely tend towards 'thoughtful creativity', exhibit a slight tendency away from 'community spirit', but lean towards being 'consultative logicians', while being slightly the opposite to 'benign acceptance'. These are, of course, the averages for a particular population of farmers.

For good management it is better to have a low score in traits 2, 3 and 5, and probably also in trait 4, whereas a high score in traits 1 and 6 may create problems.

It will be noted that some of the questions appear in more than one trait (question 16 for example). This is because some of the elements of what is embodied in a question impacts on more than one trait. Strictly it would be better to test more questions to ensure complete independence. Each question has a weighting on how much it contributes to a trait based on a statistical technique known as 'factor analysis'. For example, the weightings of the questions in trait 2 are 0.39, 0.43, 0.35, 0.56, 0.49, 0.38, 0.63 and 0.62. When naming a trait, it is necessary to take into account these weightings. Trait 2 has questions 17, 24 and 25 as being particularly important.

With respect to the five-factor model of personality, it is clear that 'correctness anxiety' is similar to neuroticism (anxiety), 'conscientious planning' to conscientiousness, 'thoughtful creativity' to openness, 'community spirit' to extroversion, 'benign acceptance' to agreeableness and 'consultative logician' relates to extroversion as well as conscientiousness and openness. This amalgam is clearly important.

Every farmer will have an unique score in the style test indicating, assuming they ranked the statements truthfully, their particular phenotype. This will impact on their managerial success in conjunction with the other important factors. It is important to understand this interaction between style and success when assessing a farmer as it helps explain the outcomes being achieved. It is common to assume relatively poor outcomes are due to inappropriate product and process selection. While this is probably true, improving the situation is often not just a matter of deciding what the selections should be, but of improving the skills of the manager so he is better able to make the selections himself. Moving in this direction starts with testing the farmer to define his management style. Astute observation of a manager as he operates adds to this assessment, though in some cases this will be all that is available. Understanding the human condition will greatly enhance the success of the observations.

Farmer Hank, one of the case studies, has an interesting managerial style. In line with his risk-averse attitude (see below) he tends towards having a moderate-to-high level of 'correctness anxiety' (score of 8 out of 20 with the lower the score the stronger the factor – the population average is 12.8) and this leads into his need for 'conscientious planning' (21 out of 40) being one of the more important style factors. Hank is constantly planning and replanning his activities with 'paper' experiments using a range of software packages. Hank tends to concentrate on the farm as a focus for his energy and is strong in 'thoughtful creativity' (10 out of 30) with many ideas bounding from his fertile mind. He also is keen on mulling over ideas with others, with his 'consultative logician' (5 out of 20) score being particularly strong. Hank is also a tolerant person and this is expressed in a similarly strong score in 'benign acceptance' (5 out of 20). Hank reckons these figures are a true reflection of his style, and the fact that he comments on his own personality suggests he is introspective and thoughtful about his own capabilities. This self-realization is an important factor in ensuring the true situation is honestly recognized.

It is interesting that Margrave, another very successful farmer, has a similar managerial style for only some of the factors. Margraves 'correctness anxiety' score is much higher at 17 out of 20, relative to the population average of 12.8. This means Margrave tends not to worry about issues, and this complies with his risk attitude as determined by the risk test given below. Margrave has a risk-neutral attitude, meaning he is prepared to take the good with the bad and seldom makes decisions based solely on their riskiness, or lack of it. Margrave is slightly less of a 'conscientious planner' (score 23 out of 40) than Hank, and this shows in the lack of detailed records he keeps, 'why keep records which I can't see a use for'. Hank might argue that you do not always know what records and information you might want in the future. When it comes to his 'thoughtful creativity', Margrave is right on the population average of 14 out of 30. A casual observer would have said that Margrave was very thoughtful and creative as shown by the wide range of activities he has instituted, so perhaps the book test was not a successful representation of his true style. Where observations question test results, further investigation is warranted.

Margrave has a strong 'community spirit', with the low score of 7 out of 20. This is the same as Hank and reflects that both farmers are involved in community activities. Margrave, for example, is involved in farmer committees that review the state of farming (in this case sheep and beef) and make policy suggestions to the administrators. Margrave is less of a 'consultative logician' than the average with a score of 11 out of 20 (population average 9.9), but has an accepting management style with a 6 out of 20 score for 'benign acceptance'. Observation backs up these formal scores.

Attitude to Risky Situations

Related to managerial style is a farmer's attitude to risky situations. A risky situation is one where outcomes cannot be predicted with certainty. Thus, for example, a crop yield can take on many values depending on, primarily, the weather and disease impacts. While the range is probably predictable, all that can be said about the yield and profit is that various levels are possible, each with a degree of probability. Many such examples abound. The more variability, the more risk. This risk can be reduced through various management techniques; an obvious one in the case of crop yields is the use of irrigation, though this may be profitable in its own right. Whatever the case, irrigation will certainly reduce the degree of risk. Farmers that invest in irrigation just to reduce the risk are referred to as risk averters. In contrast, a farmer who enjoys taking risks is called a risk preferer, though there will be all shades between the extremes. A risk preferer, for example, might well not bother with crop insurance as the possible benefits from not paying the premium more than compensates the downsides of a poor year (Fig. 2.2).

A farmer's attitude is undoubtedly a function of past experiences, his personality and the situation he finds himself in. If, for example, the mortgage repayments are high, a farmer may make conservative decisions to ensure the payments can always be met. The outcome of a poor year in which the mortgage is called in is just too horrendous to contemplate even where the farmer's basic attitude may tend towards risk preference.

Fig. 2.2. Risk attitude influences decisions. In this case storing more than a year's winter feed supply has been important.

There is nothing good or bad about a particular attitude. What is important is that a farmer is able to take note of his attitude and act accordingly. If his decisions do not match his risk attitude, it is likely that he will be concerned about his situation, which in turn will lead to a reduction in his general level of satisfaction.

This is not to say a farmer should not examine his attitude and, perhaps, make efforts to change it. An overanxious farmer is likely to be a severe risk-averter, leading to poor decisions. Making an effort to be less of an averter, perhaps with the help of a spouse and other friends, might well have a considerable pay-off in terms of achievement and satisfaction. The same comments apply to extreme risk-preferers who may well gamble away assets inappropriately.

There are a number of tests to assess a farmer's attitude (e.g. Anderson *et al.*, 1977). However, observation of everyday actions may well give a good indication of a farmer's attitude. An informal test might well be constructed to suit the particular environment. An example of this kind of test is given below in which a number of typical decision situations is offered with a range of answers each of which reflects aversion or preference. Taken together the answers lead to a conclusion.

SET OF QUESTIONS TO ASSESS A FARMER'S RISK ATTITUDE

1. Would you prefer to
    i. take out a fixed price contract on your lambs of US$3.80/kg OR
    ii. accept what the market offers at the time.

You anticipate the price will be US$3.90, but there is a good chance (40%) that it will be US$4.20, but it could be as low as US$3.20 with a 20% chance.

2. **If you have irrigation, do you have**
    i. an irrigation system that does NOT get used to its full capacity? OR
    ii. an irrigation system that IS used to its full capacity?

3. **Insurance records show that** the chance of your hay barn and its contents going up in smoke from accidental causes (in contrast to damp hay!) is only 0.1% (i.e. one chance in 1000 – would expect your hay barn to burn down once *every* 1000 years). The hay barn holds 1000 small bales (currently selling at US$5 each).

If they want to charge an annual premium of US$1,180,
WILL YOU,
    i. insure the barn and contents? OR
    ii. NOT insure the barn and its contents?

The replacement cost of the barn is US$4,000. You normally have the barn full.

4. **A wool buyer is offering you US$3.60/kg** greasy for your fleece wool, but you have a suspicion the market is going to lift. Your reading of the world scene is that the price could be as high as US$4.10 at the next sale, though as low as US$3.40 is certainly not out of the question as the world scene is somewhat shaky. In mulling over the situation you come to the conclusion that there is a 50% chance of getting close to US$4.10, a 25% chance of US$3.75 and a 25% chance of US$3.40.

WILL YOU
    i. sell to the buyer? OR
    ii. sell at the next auction?

5. **You have been contemplating increasing** your ewe numbers, as in some years you seem to have more-than-enough feed even after filling all the hay barns. The trouble is, given a series of average to bad seasons, you would struggle to feed the sheep at a reasonable level. Your calculations and hunches suggest the following:
    a. If you stay as you are – average profit per ewe will be US$52.
    b. If you increase stock numbers by 10%, profit in a typical year will be US$45/ewe, profit in a good year will be US$55/ewe, while profit in a poor year will be US$38/ewe.

Chance of a good year 30%, poor year 20%, typical year 50%.
WILL YOU
    i. stay as you are? OR
    ii. increase stock by 10%?

6. **Which best describes your betting actions?**
    i. seldom take LOTTO tickets and spend less than US$1000/year on sports betting, OR
    ii. do take LOTTO tickets more than occasionally and spend quite a lot on betting.

7. **If you rely on hay or silage for winter feeding**, do you
    i. regularly have more than 30% of your hay and/or silage left over each spring? OR
    ii. seldom have much left over?

8. If you have (had) a mortgage, do (did) you use
   i. the floating-rate option if available to you? OR
   ii. the fixed-rate option?

Skip this question if the farmer has never faced this choice.

9. For your extensive subdivision plan (perhaps imaginary) requiring fencing, improving the water supply and some track making, would you accept
   i. a fixed rate 10-year mortgage at the rate of 7.5%, OR
   ii. a floating-rate mortgage?

Your discussions with bank managers, and your reading, suggests the following: Worldwide the long-term average rate is likely to slowly decline, and it is almost certain that the average exchange rate will stay much the same as it is now, though there will be minor variations. However, history tells you that there is little that is certain about interest rates and you reckon the average rate for floating-rate mortgages could go to 8.5% with a 40% chance, but equally there is a 40% chance it could actually decline with the average over the 10 years turning out to be 6.5%. Will you take a FIXED- or FLOATING-rate mortgage?

ANSWERS

The number of times option (i) is selected relative to the number of times option (ii) is selected will tell you the person's risk attitude. If (i) is greater, the person is a risk averter, and vice versa. The degree of aversion or preference depends on the balance.

For question one, the second choice has the highest pay-off on average, but it is riskier in that it may pay less than the first choice. For question two, people not using their irrigation to full capacity, assuming this is possible, represents risk aversion in that the irrigation is being used for insurance rather than straight profit. For question three, the least-cost option is to take the risk yourself, but few would do this. Thus, buyers of insurance in this case represent risk-aversion. In the case of the wool-selling policy in question four, option (ii) is the most profitable, but is clearly riskier in that the price could be both lower and higher. A risk averter will, therefore, choose option (i). In question five, staying at the current number of animals is both more profitable and safer. However, if someone was to choose to increase stock numbers, they are very much gambling on the chance of a very good year, and are, therefore, an extreme risk-preferer.

It is clear that choosers of option (i) in question six will be risk averters as lotteries are always stacked in the favour of the organizers. In the case of the hay storage (question seven), a risk averter will always select option (i) due to its 'safeness'. In question eight, the safe and sure way (risk averter) is to select a fixed interest rate, thus giving surety; and finally, in question nine, the choices are equal in an average financial sense, but someone choosing option (i) will be looking for surety and exhibit risk-aversion tendencies.

It will be noted there are nine questions, so one of the totals for each of the options will at least dominate the other by one. The degree of domination gives the strength of the attitude.

It will be clear from these questions that, in a practical sense, careful observation of how a farmer operates will provide a quick indication of risk attitude. A consultant must weigh up whether trying to change the farmer's attitude will be worthwhile leading to greater satisfaction. In some cases, attempting change might cause more problems in that the person might become quite anxious over the whole affair. Caution is required.

The case farmers found the questions easy to answer. Discussion and observation of their systems indicated that the test results were in line with what you would expect. Hank chose option (i) six times and option (ii) thrice, indicating that he is a relatively strong risk averter, and this shows in, for example, his constant use of irrigation and reasonable feed reserves. Margrave, on the other hand, chose both options (i) and (ii) four times, with one question not answered as it was not applicable (the irrigation question). Margrave has a clear risk-neutral approach to management and, as a consequence, is content to run large numbers of animals stretching the farm's resources, but he also plans on selling off animals when the need arises in a bad year.

Despite his risk-averse attitude, and his relatively high 'correctness anxiety', Hank does not stay awake worrying about the situation if he has carefully thought through the problems and is comfortable with his conclusions. Perhaps his management style is the reason he puts so much time into researching problems and doing all the sums. Having finished his analysis, he 'rests easy', knowing he has fully investigated the situation and come to the right conclusion. Hank does, however, recognize that as a specialist dairy farmer 'all his eggs are in one basket' meaning a downturn in the world market is not compensated through diversification. Hank commented that his children have been involved in the wine industry, and he wondered about establishing a vineyard on some of the land. The initial research, however, suggested the number of degree days over the summer would probably preclude grapes having a sufficient brix reading to make good wine. Perhaps this is something for the future, once cool climate viticulture has been further explored.

Intelligence and Its Links to Management

Few would doubt the importance of intelligence in successful management. Intelligence has been studied by many over a long period of time, yet there are still many opinions on exactly what is meant by intelligence. Early definitions specify intelligence as the 'capacity to learn from experience and the ability to adapt to the surrounding environment' (see Sternberg, 1995). Generally speaking, most popular belief on intelligence regards it as the ability to understand situations and ideas as well as sort out problems, and also having an extensive store of information that is readily acquired. That is, being logical and having a good memory. Clearly these two components are critical to good management. Personality, on the other hand, would commonly be regarded as the emotional and people aspect to being human.

In that there are many components to human lives, intelligence has many aspects. Gardner (1993) lists the components as:

- verbal–linguistic   i.e. the ability to use words and language;
- logical–mathematical   i.e. the capacity for inductive and deductive thinking and reasoning, as well as the use of numbers and the recognition of abstract patterns;
- visual–spatial   i.e. the ability to visualize objects and spatial dimensions, and internal images and pictures;
- bodily–kinaesthetic   i.e. the wisdom of the body and the ability to control physical motion;
- musical–rhythmic   i.e. the ability to recognize tonal patterns and sounds, as well as a sensitivity to rhythms and beats;
- interpersonal   i.e. the capacity for person-to-person communications and relationships; and
- intrapersonal   i.e. the spiritual, inner states of being, self-reflection, and awareness.

Clearly some of these so-called independent intelligence traits do not have a bearing on management. Of course, other psychologists would not totally agree on this list.

In a general sense, intelligence is often divided into two:

- fluid; and
- crystallized.

Fluid intelligence covers broad basic reasoning, which is largely genetically sourced, whereas crystallized is fluid intelligence as expressed in a particular culture. Adjectives that have been used to define the components of fluid intelligence are:

- inference;
- induction;
- memory span;
- intellectual speed;
- visualization; and
- retrieval capacity.

For crystallized intelligence, words used include:

- verbal;
- mechanical;
- numerical; and
- social skills.

If anything, there is more consensus over what is called the Triarchic Theory of Human Intelligence than most concepts. This was proposed by Sternberg (1995) and has three basic components described as the cornerstones of a triangle. These are a person's

- analytical abilities;
- creative abilities; and
- practical abilities.

Adjectives that come to mind for each apex of the triangle are:

- analyse, compare and evaluate;
- create, invent and design; and
- apply, use and utilize.

In a management sense, the basic three components are all relevant, as are the facets of each attribute. Successfully analysing alternative causes of action is a critical skill in management, as is creativity in that it is critical that a manager can create solutions to problems in the widest sense of the 'problem' word (a problem might well be the question of what is the best set of crops to grow next year, for example. And then there are the literal problems such as what to do if the stock feed is running out). In addition, of course, practical ability is an absolutely necessary skill in primary production in that success depends on physically getting the farm and its animals into an appropriate state. While a manager might employ people to carry out the physical tasks, it is not possible to plan, organize and control outcomes unless a good understanding exists.

It has been noted that experience of the right kind is probably an important aspect of managerial ability, as it is in many occupations. Learning from experience is not just a matter of existing through various situations, but a matter of:

- observing the situation;
- analysing it; and
- learning the lessons available for future reference.

Thus, high intelligence succeeds in obtaining the most from experience, and of making use of the stored lessons in solving future problems. As noted earlier, the making of an expert involves learning from experience.

This brings up the memory aspects of intelligence. Memory is generally regarded as limitless, though you may not agree when it comes to recalling the things you have learnt. It is a matter of training yourself to remember and recall items at the correct time. Research, and common observation, suggests that memory is divided into short- and long-term components. You constantly observe sights, sounds, smells and feelings, and these move into short-term memory where the inputted observations are processed for meaning and relevance. Much of the data is immediately discarded. Think back to 5 min ago ... do you still remember the sounds and objects you observed? Much of the detail has been lost, or at least not recallable. If, on the other hand, you heard, for example, a violent shutting of a door you probably still have this in mind. The reason is that the short-term memory and the processing system noted this auditory input as being unusual, and so it was passed into long-term memory for later recall. Essentially, the processing filters material to assess whether it is valuable, and if so pushes it through to long-term memory, where it can be stored, perhaps for ever. People who have suffered physical brain injuries usually comment that they cannot remember the knock and what went before it. This happens as the blow interrupts the processing stage and so the conclusion is not sent through to long-term memory. The lesson from this situation is that a good manager will train to ensure the appropriate processing occurs with the useful material being shunted through to long-term memory. Good training will ensure that the crystallized intelligence is well honed for the management task.

Short-term memory and the processing of material entering are thought to be limited. However, this probably varies with a person's level of intelligence in that we all know that some people are better at observation, and sorting out what is important and the relationships between the observed information. You will be aware that telephone numbers are usually limited to seven digits. Tests suggest seven is the maximum number of items fitting into immediate short-term memory (five to nine items seems to be the range of the limits). If too much information arrives too quickly, a considerable quantity is lost unless the person has very high processing speed (thus speed is considered part of high intelligence). In agriculture, the speed skill is probably not as critical as in other industries in that action does not usually have to occur instantly, it is possible to observe and consider before action in a day or two.

The skill of committing material to long-term memory usually involves rehearsal, thus giving a strong imprint. The more rehearsal, the better. With practice and a knowledge of how to rehearse long-term memory, storage and retrieval can be enhanced considerably. In effect, intelligence can be improved, particularly the crystallized aspects.

It has been mentioned that intuition, or tacit knowledge, is important. In an intelligence sense, as was noted for experts, high intelligence assists this process of taking into long-term memory a store of helpful tips that creates a successful intuition. Kelly (1992) talked about the concept of holding decision rules through what he called 'constructs'. The totality of a person's constructs make up what can be called 'intuition'. Kelly believed in 'man as a scientist' in that 'man' was always striving to sort out rules by which he could operate. Thus observation led to the idea that people invoke a particular construct that suits the situation, that is, a decision on what action to follow. With experience, the constructs became better informed and appropriate to each individual. If the person is uncomfortable with a construct, searching occurs until the decision maker finds one acceptable to his psyche. Such constructs might relate to aspects of a person's personal life, or in our case, their professional life. When you analyse how you make decisions, it is seldom a full and clear analysis in contrast to a fairly rapidly 'arrived at' conclusion, you have brought out a construct. Clearly, a manager's intelligence level will impact on the process of creating suitable constructs (sometimes called 'rules of thumb', or heuristics).

What parts of intelligence are important to management? Logic would suggest that a good memory of retrievable material of relevance to primary production is a starting point. Also important is the ability to learn from experience. It is not possible to appreciate the important aspects of primary production, marketing, labour selection and management as well as financial organization without actually being involved. The term used in days gone by was 'the university of hard knocks'. And it is true that no amount of book knowledge will enable a manager to determine the important aspects of successful management. Consequently, the manager who can learn the 'constructs' necessary from practical observation and experience will more likely succeed than others who seldom learn from mistakes and experiences.

The third component is creativity. Problems frequently require innovative solutions, and certainly opportunities need observing and exploited using, possibly, new production processes that suit the particular farm. Creativity is related to imagination and both these attributes are important, provided the results of the creativity can be turned into practical solutions. Practicality is an important attribute of intelligence where production is biologically based. There is nothing mechanical nor predestined about it, so textbook solutions are seldom appropriate in a rote sense.

Moreover, an ability to understand the economic principles and decision rules relevant to agricultural production is critical, as is a knowledge of how to calculate the value of the objectives as a result of some proposed decision and change. Generally this attribute might be classed as skill in farm-based calculations. This does not mean being a maths guru, but knowing what items to include in a simple sum, and how to summarize the result. For example, given an average death rate for a mob of animals, and the fertility levels, and the number that need culling out for poor production characteristics, how many must be retained each year in order, say, to increase the number of animals by 5%. And another example might relate to the profitability of devoting a fixed proportion of the resources to a new enterprise such as a cut-flower crop. What is the lost net income from the resources diverted in relation to the increase in income from the new crop? This involves many simple estimates and calculations involving, for example, the labour freed up, the labour required, the timing of all these events (labour cannot be stored), and so on for all the other resources such as the machinery, the working capital, etc. Generally, this whole area of economic principles and simple farm sums is part of general logic skills, as are other components of good management.

Finally, given the physical nature of farming, an ability to envisage shapes is another aspect of intelligence that is desirable. For example, on a hill country farm, the plan might be to create some more fields by subdividing existing areas. In planning this it is important the farmer can work from a map to estimate fence lines and order the appropriate material. Another example is an ability to assess animals and their condition. Dairy cows need to be at least a certain live weight to produce well, and given the research that relates body shape to live weight, a farmer needs to be able to judge animals using these relationships. Another example involves pastures. With practice it is possible to judge the dry matter of a field from simple eye observation, so that this judgement together with a knowledge of the species and their stage of development enables an assessment of the energy and protein available. Again, this ability to assess the physical observations and their significance is a valuable attribute in a production system where it is not possible to be constantly reverting to expensive meters.

In summary, the important attributes that relate to intelligence are:

- memory;
- an ability to constantly improve the store of successful 'constructs' created through experience;
- creativity (imagination);
- calculational ability in its widest sense; and
- an ability to work with shapes.

General logic is also important. Working out just how good a particular farmer is with respect to each attribute can come from observation as the farmer carries out day-to-day activities, or through a formal test, or a combination.

Intelligence and Tests

There are a wide range of intelligence tests available. Generally, most farmers are somewhat dubious about being asked to sit for such tests. Furthermore, there is no general data linking the results of general tests to managerial skill, though you would expect this to be the case. There is, however, data that links intelligence tests to school results, and also to tertiary study. Consequently, educational results can give a good idea of general intelligence as defined for the purpose of the tests. Similarly, astute observation of a farmer when working closely with them on, say, setting up plans and courses of action, will give a good idea of what might be called 'farming intelligence'. And then, as outcomes are observed, and similarly, the thinking process a farmer uses, such estimates become refined.

Alternatively, or perhaps in conjunction with, tests designed for the intelligence attributes important to primary production might be created and used. Such tests need to be set up for the specific environment and production type relevant to the farmers of interest. It is necessary to use questions that relate to the particular farming type, though some questions can be universal. For example, it is no use asking a glasshouse farmer questions that relate to animal calculations as they will be unfamiliar with the relationships involved. Questions on simple logic, however, can be universal.

The abilities that cannot be tested are practical skills and understanding, so judgements in this area must rely on observations of past outcomes and current activities.

Listed in Appendix 2B is an example of a test used for broadacre sheep farming. It will be clear that many of the questions would need changing for, say, dairy farmers. The test is referred to as a 'managerial aptitude test' in contrast to intelligence, as it is this skill which is being assessed.

To score the test, the number of correct answers is added up. For some questions the answer is simply right or wrong, but for others a degree of truth exists. In this case, a suitable scoring system template is necessary. Things like creativity are very hard to test for, and in this case scores were based on the number of realistic ideas produced. Scoring the lessons from experience is also difficult, so, similarly, the number of lessons that were realistic provided the score. For calculations, an answer within a reasonable range of the true answer was accepted as correct. It will also be noted that the section labelled general was largely used to assess simple logic.

This test was given to a large number of farmers of all types, with results from 490 being included. The scoring was normalized to provide an average score of 100 in the same way as standard intelligence tests. The distribution of results is provided in Fig. 2.3.

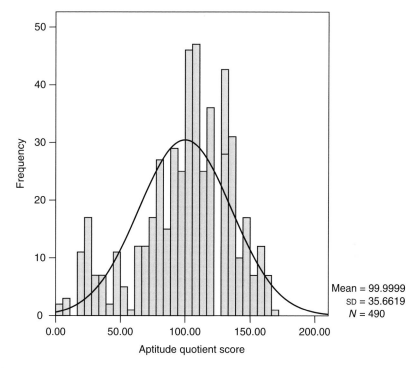

Fig. 2.3. Aptitude quotient distribution frequency.

The average of 100 will be noted, and the distribution is similar to normality as shown by the dark line. It was found that the score was significantly correlated with age, education level (years of schooling or tertiary study) and the average per cent score attained in the last year of formal study. The aptitude quotient was predicted by the equation:

$$AQ = 87.5 + 7.0\,E + 0.1\,P - 0.5\,F$$

Where E is the education level based on a score of 1–5, where 1 is primary schooling only, through to 5 for 2 or more years of tertiary education. P was the average percentage grade for their last year of formal education, and F was a code reflecting the type of farming. The equation and individual variables were all significant.

A good question is whether using these kinds of tests gives a better idea of aptitude than simply using knowledge of the farmer's education background. The answer is mixed. The predicting power of the equation was 27%. Thus, 27% of the variability is explained by the equation, leaving 73% due to other reasons. Thus, while the education background is useful in assessing a farmer's ability, many other factors, perhaps experience, are also important. This is borne out by the fact that some very good farmers do not have much formal education. Also, it will be noted that the number of years of education is rather more important than the final-year grade.

Knowing a farmer's aptitude is important when considering how managerial ability might be improved. The aptitude score clearly indicates the possibilities with respect to the potential to manage complex operations. However, the manager is a total package and while aptitude might be limiting, this may well be compensated for in part by a very suitable personality. Thus, when looking at what might be a suitable strategic plan both aspects must be taken into account.

Furthermore, when providing assistance to a manager, a knowledge of the farmer's skills within all the important aspects of intelligence indicates those areas where most help is needed. For example, perhaps the farmer is not particularly good at calculations, therefore help in sorting these aspects will be beneficial.

While the genetic side of aptitude is fixed, from the description of intelligence it is clear improvement is possible. That is what the education system is all about, but in this case we are dealing with people in the workforce, so changes either come about by organized educational programmes, or self-education. In most countries, distance learning is available so given a strong motivation there are opportunities to tackle improvement. An extension group, or consultants, can similarly organize suitable programmes, perhaps on a one-to-one basis in some cases. Where a farmer seeks help from a professional, they can consider taking extra time to explain and train in contrast to simply providing a fixed plan or action.

When given the aptitude test, the case farmers did find answering was a challenge, but given the situation they were determined to take a reasonable time for fear of being thought of as unintelligent. They need not have worried. One obtained a score of 154% and the other 174%. These farmers are clearly well into the top section of the distribution. Both have some tertiary education so perhaps this is one of the major contributors, though their background and personality will also influence the motivation to take and succeed in tertiary establishments. Certainly Margrave maintained that while he had attended very many short courses, where he gained most was from the tertiary education which forced him to develop his problem solving and analytical skills. He also mentioned the importance of being generally critical in a positive sense, and that this skill was very much a product of his tertiary education. As an aside, it is interesting to note that the survey work referred to in creating the aptitude distribution, and also other survey work, indicates the amount of tertiary education found among farmers is much more than that in the general population (around 30%).

Motivation, Objectives and Emotional Intelligence

You would imagine a farmer's motivation and objectives would influence managerial skill. A motivated farmer will seek to be good at his job, and seek out ways to improve. And his objectives may well influence the shape of decisions made, and therefore impact on outcomes. In assessing a farmer, it is important to know his objectives as someone not primarily motivated to achieve, say, maximum profit, will operate on a different plane to someone who is.

For example, Margrave, one of the case study farmers, noted he had 'unfinished business' when he left school, for while his early life was on a farm, he had not been involved other than carrying out physical chores. His family were almost secretive about business and decision making. Margrave believes his interest in agriculture was almost 'innate' with the passionate interest being 'in my genes'. Margrave could not wait to sign up for short courses and get to work on farms. Located within easy distance was a Ministry of Agriculture training establishment that constantly offered 'live-in' short courses. Margrave enjoyed them immensely and absorbed the lessons in an almost sponge-like fashion. This real passion gave rise to an extremely high motivation, which drove him to learn and learn despite long hours and exhaustion from the physical farm work. In the end, Margrave rated himself as being 'quite good' as a manager, and noted that the farm had increased in animal capacity by over 6000 ewes in recent years. You cannot doubt his skill and motivation.

Where farmers have different objectives, can it be decided which are more efficient and managerially expert? Such an assessment, of course, depends on the objectives so a farmer who wishes to, say, operate an organic farm can be just as managerially efficient as a farmer maximizing profit. An observer of management must therefore take into account the 'farm's' objectives, as well as the motivation to succeed. In some cases, it will be possible to observe motivation, and consider whether inappropriate motivation is getting in the way of efficiency. Changing a farmer's level of motivation, on the other hand, might be more difficult.

This section contains a discussion on motivation, its relationship to personality (managerial style), and considers how objectives might be quantified remembering that many farmers are owner–operators so determining objectives is easier than if the ownership is complex and divorced from the farm. Also discussed is the psychometric test Emotional Intelligence (EI) as some believe that what it embodies relates to managerial outcomes. As the name suggests, it is to do with emotions related to getting 'things' done.

There are many theories on motivation: what it is, and how it might be measured. Furthermore, the jury is still out on whether it is a separate human trait, or whether it is an expression of personality and a sub-component reflecting the basic traits. However, it is an obvious trait to discuss when considering managerial skill and, therefore, must be included here.

Most people would consider 'motivation' to involve not only an ability to think of, and initiate, a project, but also the control of the direction of its conduct and final completion, including persistence to finish off what is started, if this is indeed rational. Further, motivation probably relates to a 'need, want or interest' a person has that propels them in a certain direction. Frequently the motivation to achieve a goal will stem from 'intrinsic motivation' meaning that the drive comes from the pleasure it brings, or from a feeling that the task is important, or that it is morally required. All these drives are internal with success leading to the person feeling better about themselves. In contrast, the drive to achieve might be 'extrinsic' meaning it comes from outside the person. An example would be the bank manager demanding repayment of an overdraft within a defined period.

Self-control is an important part of motivation. This refers to the ability to perform the tasks required to achieve the goal or need in a timely manner. Many people have the right aims but actually carrying out the tasks on time is another matter and requires self-control. Sometimes the degree of self-control depends on the needs. We all need to eat, but few in the Western world find it difficult to raise the self-control needed to achieve basic bodily needs, indeed the control required is in not eating too much. As the 'needs' become more divorced from these basic needs of life and society, it often becomes more difficult to carry out what is required. Thus, the motivation to become a perfect manager involves more and more self-control and dedication.

There are theories on what drives people to achieve their goals. What drives you to study? Perhaps it is your conscience that will not be content until you have achieved a goal. You have an internal drive that forces you to continue until your system reaches a comfortable equilibrium. Or perhaps you have some external incentive to drive you on to success. An example might be a promise of a substantial reward once you have achieved success. Clearly your emotions are involved in the drives. Emotions give rise to a state somewhere on the continuum of unpleasantness–pleasantness, so if you are feeling uncomfortable about a situation you may well strive to fix the problem and move yourself into a state of pleasure. Just how much effort goes into creating this change will depend on the degree of unpleasantness and your means to change the situation. If you have no control over the situation, you may change your feeling as you rationalize that it is impossible to do anything. Some people, however, while realizing it is not possible to alter the situation, still find it unpleasant and consequently loose sleep. You might class such people as 'anxious'.

Emotions can also cause physiological responses such as a rise in your heart rate and associated panic reactions. On the positive side, feelings of achievement might cause relaxation and a balanced bodily state that even perhaps leads to good health. In a farming sense, such emotional reactions can stimulate appropriate action such as getting out of bed in the middle of the night to attend to a problem such as potential flooding.

One wonders how there can be such a variation in motivation within society. It has been proposed that there are major differences between individuals with respect to their need for:

- achievement;
- power; and
- affiliation.

You might also add to this the need for praise and positive feedback. It seems factors that encourage motivation are:

- autonomy;
- competence feedback; and
- relatedness.

It is likely that the time between action and outcome is also a relevant factor in that if action and response occur close together and it is an easily observed outcome, the motivation to achieve and complete a job is higher. We are all aware of this, particularly when studying for a qualification that might be years away.

In a farming sense you can observe people who are very proud of their achievements, and others who just simply 'get on with the job' with little fuss or fanfare. The personal power factor is somewhat less important in that it is only in large farm situations where considerable labour is employed giving the opportunity to feel powerful, though in some cases the 'power' is felt through being able to control a large area with many animals. The need for affiliation is often expressed through joining, mainly, rural organizations. It is interesting to note that this need is likely to be related to the extroversion trait.

In a practical sense, expressions of motivation can be seen in how people organize themselves. Some managers make lists of prioritized jobs, and as they are achieved they get crossed off leading to satisfaction that is clearly noted. Such lists can be reordered as conditions and situations change. This process is part of self-created positive feedback and helps maintain motivation. Similarly, a good record-keeping system helps observe where achievements have occurred, particularly with respect to physical outputs relative to inputs and, thus, efficiency changes. Farm accounts, if properly kept, will also lead to noting changes in profit, and provide feedback both in a motivational and also in an economic sense.

Of course, talking about efficiency suggests that humans have a tendency towards rationality. However, while most would say they strive for rationality, in practice straying from rational decisions often occurs. No doubt this relates to the strength of a farmer's motivation relative to the goals. Sometimes emotion takes the better of rational thought, leading to erratic decision making that may depend on the feelings on the day. This is where a conflict of motives plays out, so that, for example, the profit motive loses out to the enjoyment, say, of seeing perfectly presented fences round a field, or perhaps the draw of a family reunion in contrast to harvesting a crop on time. Humans are complex organisms ruled by a myriad of factors as this discussion has pointed out. It is the play of all these factors that leads to an individual's motivation.

While there are many tests for personality and intelligence, few exist for assessing motivation. This means astute observation is the main tool. Over time, a clear feel for both the farmer's motivation and the relevant goals will surface. Attention to detail, degree of careful planning, determination to get the jobs organized and achieved on time are all indicators as is the concern for exploring all new and possible technologies and products. Does the farm come first where conflicts for time and energy exist?

One test that has been explored is called the Motivational Trait Questionnaire (MTQ; for reference see Kanfer and Ackerman, 2000). This test is not specifically designed for agriculture. It is based on the theory that motivation is made up of three factors:

- personal mastery;
- competitive excellence; and
- motivation related to anxiety.

Each has several facets. Personal mastery is made up from:

- the desire to learn; and
- mastery (e.g. 'I set high standards for myself and work toward achieving them').

Competitive excellence has the following facets:

- other referenced goals – this relates to whether you compare yourself to others' achievement (e.g. how does your milk production per hectare compare?); and
- competitiveness (e.g. 'I would rather cooperate than compete' (reversed scored)).

Finally, motivation-related anxiety is made up of:

- worry (e.g. 'Before starting an important project, I worry about the consequences of failing'); and
- emotionality which records a person's emotions related to outcomes.

The big question is whether motivation measurement does in fact tell you anything more about a farmer than the results of the management-style test, or similar personality tests. From the discussion you might conclude that the general anxiety trait and the conscientiousness trait might well record most of what there is to know about a person's motivation. Further, the extroversion trait is likely to relate to feedback information through contacts with other people. Indeed, whatever research that is available does suggest there is a strong correlation between motivation and personality such that separate motivation tests add little information. Thus the comments above about astute observation... Given personality test results and observations it should be clear what level of motivation is likely.

In the case study farmer Hank's background there may be specific factors that created the high level of motivation he clearly has. This is in contrast to the idea that general management style is all there is to know about a person. In reality motivation, in some cases, will be due both to their genes and their general experiences as well as any notable and especially memorable experiences. Hank recounts how he was heavily involved in the local agricultural club in which, as a child, you were encouraged to take part in a range of competitions such as preparing and showing your pet animals (calves in his case). Hank could not bear to lose; he notes 'I'm competitive. A failure is a great incentive'. He also remembers missing out on achieving at football and became quite upset as with his competitive nature you had to win. This competitive nature has carried over to adult life and provides a strong motivation to succeed. Hank even goes as far as to note 'I had quite a problem of pride in my boyhood. It was cause for upset, but also determination'. This is despite his parents 'not being as driven as I am', suggesting that his early experiences were the seed for his strong motivation. But, of course, his personality would have affected his feelings over failures for others might have just brushed over the events with hardly a touch of downcast feelings.

Examples of research relating motivation to personality include Judge and Ilies (2002), who found that the basic five traits had a multiple correlation with motivation of 49%; Kanfer and Ackerman (2000), who concluded that anxiety was highly correlated with the motivation traits, but motivation was not related to either fluid or crystallized intelligence and is, therefore, independent of intelligence. In addition, Zuckerman *et al.* (1999) found correlations as high as 80% for motivation and the big five personality traits. They concluded that anxiety and extraversion were important traits in motivation.

In view of these results, it is believed that developing motivation tests specifically for agriculture will not provide additional useful information about a farmer over and above management-style test results combined with astute observation.

In contrast, however, there may well be value in conducting a formal assessment of a farmer's objectives. The results can be combined with the observations that anyone dealing with a farmer will inevitably make as they reflect on the decisions made and the objectives they portray. The next chapter contains a set of questions that can be used to quantify a farmer's objectives, or indeed the objectives of other members of the family and management team.

While a farmer's motivation is highly related to his personality, the objectives held are probably very dependent on the influence of the farmer's family, particularly in his earlier years, as well as the influence of the community within which the farmer grew up including the effect of the paradigms expressed during schooling. Just what gave rise to the final motivation level and objectives is probably complex and highly unique for each farmer. What is important, however, is making a judgement on the level of motivation held by a farmer, and the objectives and goals driving the farmer. Any professional should discuss the situation with a farmer with a view to assessing the appropriateness of the goals and encourage change where this is clearly what the farmer and the family want. Similarly, where motivation is a problem, counselling and support may well be appropriate in the interests of improving managerial ability and efficiency. (For further information on objectives, and a set of statements that can be used to assess a farmer's objectives, see Chapter 3.)

While dealing with the personal factors of motivation, objectives and goals it is also useful to comment on what has become known as EI. This is a recent concept and covers intelligence as related to the emotions in contrast to intelligence related to the aspects normally tested in IQ tests (reasoning, memory and so on). Some would say, of course, that these aspects of a person are measured through personality tests.

Emotional intelligence (Zeidner *et al.*, 2004) is made up of:

- awareness of emotions in self;
- awareness of emotions in others;
- management of emotions in self; and
- management of emotions in others.

Clearly, a manager skilled in all these aspects is likely to achieve better outcomes than those not similarly skilled, especially where the farm employs a reasonably sized workforce, and deals with many contractors. It is suggested that EI is related to:

- effective networking;
- conflict management;

- stress management and adaptability;
- negotiations; and also
- listening and verbal communication.

All of these attributes are regarded as desirable in good management. Psychometric tests are available for EI and follow the usual idea of a series of questions which are then scored.

The tests are not presented here for it seems that EI is highly correlated with personality. One study concludes that the Myers–Briggs personality trait 'intuition' is significantly and positively related to higher levels of EI (Higgs, 2001). Zeidner *et al.* (2004, p. 384) also conclude that 'despite the important role attributed to a wide array of emotional competencies in the work place, there is currently only a modicum of research supporting the meaningful role attributed to EI'. Given these conclusions further discussion of EI is not considered beneficial at this stage.

Learning Style

The final test to be presented relates to assessing a farmer's learning style. Good managerial skill involves constant learning, both in a skill ability sense, but also ensuring a farmer's technological skill base is contemporary, as should be his knowledge of markets and regulations. Farmers must be constantly learning. Just how well such lessons are absorbed will depend in part on how they are presented as we all respond best to material presented in a way that suits us.

There are at least 70 or so tests that have been developed to assess a person's learning style, and each one has an associated theory. It does seem that few of these have any scientific backing so there is little general agreement on the learning process and how it might be assessed. However, about 12 of these tests do get used. One common background idea is that people learn in four basic ways:

- visual (learning by seeing);
- verbal or auditory (learning by hearing);
- reading or writing (learning through processing text);
- kinaesthetic (learning by doing).

In reality, it would be surprising if we do not learn by many of these methods, but some will suit more than others in each particular situation and problem.

One test that has been used quite extensively was developed by Kolb (1984) and has subsequently been upgraded several times with the latest upgrade in 2005 (Kolb, 2005). This test has been extensively researched with considerable technical support available for interpretation.

Kolb believed that the learning process involved:

- having a concrete experience;
- this leads on to observation and reflection;
- giving rise to abstract concepts which are mentally held;
- these are then used in new situations to provide a test of the concept.

If the concept does not work, then the *whole process starts again* until a match between the concrete experience and the concept is arrived at.

Just how important each aspect is to an individual will vary so that, for example, they might find the concrete experience very important in contrast to, perhaps, being able to work from more abstract representations such as a book description of a problem and solution. Thus, the idea of learning styles was developed. While there is a continuum, four distinct classifications were developed. These are (reprinted with permission. For details see the acknowledgements on p. xi. Source Kolb Learning Style Inventory):

The diverging style

This combines the Concrete Experience and Reflective Observation phases. People with this learning style are best at viewing concrete situations from many different points of view. Their approach to situations is to observe rather than take action. If this is your style, you may enjoy situations that call for generating a wide range of ideas, such as brainstorming sessions. You probably have broad cultural interests and like to gather information. In formal learning situations, you may prefer working in groups to gather information, listening with an open mind and receiving personalized feedback.

The assimilating style

The assimilating style combines the Reflective Observation and Abstract Conceptualization phases. People with this learning style are best at understanding a wide range of information and putting it into concise, logical form. If this is your learning style, you probably are less focused on people and more interested in abstract ideas and concepts. Generally, people with this learning style find it more important that a theory have logical soundness than practical value. In formal learning situations, you may prefer lectures, readings, exploring analytical models and having time to think things through on your own.

The converging style

This style combines the Abstract Conceptualization and Active Experimentation phases. People with this learning style are best at finding practical uses for ideas and theories. If this is your preferred learning style, you have the ability to solve problems and make decisions based on finding solutions to questions or problems. You would rather deal with technical tasks and problems than with social and interpersonal issues. In formal learning situations, you may prefer experimenting with new ideas, simulations, laboratory assignments and practical applications.

The accommodating style

The accommodating style combines the active experimentation and concrete experience phases. People with this learning style have the ability to learn primarily from 'hands-on' experience. If this is your style, you probably enjoy carrying out plans and involving yourself in new and challenging experiences. Your tendency may be to act on intuition rather than on logical analysis. In solving problems, you may rely more heavily on people for information than on your own technical analysis. In formal learning situations, you may prefer to work with others to get assignments done, to set goals, to do fieldwork and to test out different approaches to completing a project.

For simplicity, the four types can be abbreviated to concrete experience (CE), reflective observation (RO), abstract conceptualization (AC), and active experimentation (AE).

The four learning modes can be combined into two 'scores' that result from combining concrete experience and abstract conceptualization (AC–CE), and combining reflective observation and active experimentation (AE–RO). Parameter (AC–CE) measures the extent to which a person emphasizes abstractness over concreteness, while parameter (AE–RO) measures the relative 'action-over-reflection' emphasis. By combining both parameters, a two-dimensional space is developed, and the four learning styles are further defined as:

- convergent, which emphasizes abstract conceptualization and active experimentation;
- divergent, which emphasizes concrete experience and reflective observation;
- assimilation, which emphasizes abstract conceptualization and reflective observation; and
- accommodative, which emphasizes concrete experience and active experimentation.

Following Kolb's system, to work out a farmer's learning style requires obtaining his ranking of a series of 12 adjective sets. For each set, the farmer must rank them in degrees of correctness for his particular approach, or view of life.

For example, for the adjectives 'inquisitive', 'bored', 'unadventurous' and 'excitable', a farmer might decide 'excited' is very much like him and so is scored 4, whereas he is certainly not 'bored' so this adjective is scored 1, and similarly 'unadventurous' does not normally apply to him so it is scored 2, leaving 3 for 'inquisitive' which he certainly is.

By way of an example, a group of dairy farmers in New Zealand scored such that 23% were accommodators, 20% assimilators, 23% convergers and, therefore, 34% were divergers. As you would expect, the people requiring concrete experiences for learning dominate. Divergers also use 'reflective observation' in contrast to experimentation, which is perhaps a little surprising for farmers. Of course, just how a person learns will not necessarily impact on their managerial ability, but a knowledge of the best way for an individual to learn helps in working on developing a manager's skills. Thus, while concrete experience is dominant, there are still people who can learn in abstract form so that they can improve from reading. Concrete-oriented people, however, will best be helped through demonstrations and practice, though in the end everyone does need to have hands-on experience.

It should also be noted that the evidence relating learning modes with outcomes is not particularly strong. However, common sense would suggest that some people do learn more efficiently through certain modes than others. For the case study farmers, Margrave is very much an active experimenter/concrete experience person with an AE score of 30 out of a possible maximum of 36, and CE 30/36. In contrast, his abstract conceptualization (AC) score was 12 and reflective observation (RO) 18, indicating that he 'learns by doing'. For Hank, his scores are much more even across the types with CE 26, AE 24, AC 21 and RO 19. The fact that Hank spends at least an hour a day reading complies with his more abstract learning style compared with Margrave, though he still does score reasonably highly on the 'doing' categories.

Whatever research is available on learning is largely non-agricultural so it cannot be categorically concluded that, for example, the Kolb test will be useful. It could be that specialist tests designed for primary production might be more useful. Some farmers will find interpreting the adjectives in the Kolb test difficult so that a test using similar terminology that can be directly related to agriculture might well be more useful.

In the Kolb concept, experience is an important part of the learning cycle. Even if theorists do not agree with the approach, all would agree that experience and practice is an important part of becoming a good manager. The ability to observe, learn, and implement the lessons from experience is probably very important in such a practical occupation as primary production. More is said about experience in the next chapter.

Margrave, the case study farmer, has an interesting learning style. He firmly believes 'reading is the style I learn best from. I read everything I can, particularly during meals when I was young and single. The reading led to thinking of options...'. Hank, in contrast, believes he is more of a kinaesthetic learner and needs to get out and practice. He found learning computer skills needed plenty of time at the keyboard in contrast to reading manuals. However, Hank also noted his memory was very good, something everyone aspires to, and for this reason makes sure he reads at least an hour per day to build up his knowledge and obtain ideas as well as critique others. 'There is so much to cover I have little time to do other early morning and late night activities. I haven't read a novel in 20 years'.

While the discussion has focused on the Kolb (2005) test, as noted there are many learning style tests available. Most involve answering a set of questions which are then scored leading to a conclusion using results from many uses of the tests. Two examples of the kinds of questions are given below (a search on the World Wide Web (WWW) will provide a reader with many tests. Before use their validity should be checked through the literature; the examples provided were picked at random).

For these example questions each is scored on a 1 to 4 scale of correctness.

I feel the best way to remember something is to picture it in my head.
I follow oral directions better than written ones.
I often would rather listen to a lecture than read the material in a text book
I am constantly fidgeting (e.g. tapping pen, playing with keys in my pocket)
  (see www.ldpride.net/learningstyles.MI.htm (4/09) for the full test).

And another set of examples is:

I understand something better after I
(a) try it out.
(b) think it through.

I would rather be considered
(a) realistic.
(b) innovative.

When I think about what I did yesterday, I am most likely to get
(a) a picture.
(b) words.
  (see www.engr.ncsu.edu/learningstyles.ilsweb.html (4/09) for the full test)

Clearly the scoring system for each of these examples is different. Thus, the interpretation will depend on the scores obtained relative to a standard set.

Concluding Comments

Thus, the age-old question is – 'is an excellent manager born or trained?'. To answer this question, it is important to understand the components of a good manager, and to consider whether the components can be improved through training of some kind. This chapter has contained discussions on the factors likely to impact on good management, and provided formal written tests that can be used to assess the characteristics of farmers. The results enable us to form an opinion on the attributes that relate to good management. The tests cover managerial style (personality) and aptitude (intelligence). Also discussed is a farmer's attitude to risky situations as this will influence the form of decisions made, as will a farmer's set of objectives. Furthermore, as the effort a farmer puts into making good decisions is related to his belief in just how much control he has over farming outcomes, a test for assessing his 'locus of control' will be introduced in the next chapter. Finally, as the objective of studying managerial skill is to improve farmers' skill level, a test of a farmer's learning style was discussed. Some doubt about its efficacy and value, however, was commented on. Similar comments apply to motivation tests, and the concept of 'emotional intelligence'.

To understand skill improvement, it is important to understand the basic characteristics that give rise to skill. The personal traits discussed are the building blocks, so familiarity with them leads to a sound grounding for understanding people and their abilities. This chapter has provided this background and understanding. Another characteristic that has not been mentioned is 'entrepreneurship'. This term is used to describe people who create new ideas for products and production systems, and are able to put the ideas into successful operation. Whether the basic personality traits are highly related to this characteristic is not yet clear, but this is certainly a strong possibility. The openness trait could well give rise to entrepreneurship. Entrepreneurs might also be classed as 'experts', as discussed earlier. However, further research may more directly correlate entrepreneurship with personality, or conclude whether a new trait needs to be added to the management-style traits.

Finally, it is worth reinforcing the comments about the use of normal observation in classifying managers, and assessing their current rating in the skill continuum. Careful and continuous observation of how a manager operates should give an astute observer a good representation of a manager's characteristics, especially an observer who fully understands the traits and factors that give rise to high skill. Where formal tests can be an advantage is for people who do not have these judging abilities, and, second, as a discussion point when working on improving a farmer's skill level. In the end, a combination of tests and observation will be important in most cases.

Chapter 3 looks at quantifying the importance of the factors contributing to managerial skill. With this knowledge it is possible to judge where any improvement effort should be focused.

Appendix 2A. Managerial Style Test

Tick ONE box that best records your degree of belief in the statements.

1. You tend to mull over decisions before acting.   TRUE ❑ ❑ ❑ ❑ ❑ NOT TRUE

2. You find it easy to ring up strangers to
   find out technical information.   TRUE ❑ ❑ ❑ ❑ ❑ NOT TRUE

3. For most things you seek the views of many
   people before making changes to
   your operations.   TRUE ❑ ❑ ❑ ❑ ❑ NOT TRUE

4. You usually find discussing everything with
   members of your family and/or colleagues
   very helpful.   TRUE ❑ ❑ ❑ ❑ ❑ NOT TRUE

5. Where there are too many jobs for the time
   available you sometimes become
   quite anxious.   TRUE ❑ ❑ ❑ ❑ ❑ NOT TRUE

6. You tend to tolerate mistakes and accidents
   that occur with employees and/or
   contractors.   TRUE ❑ ❑ ❑ ❑ ❑ NOT TRUE

7. You share your successes and failures with
   neighbours.   TRUE ❑ ❑ ❑ ❑ ❑ NOT TRUE

8. Keeping records on just about everything is
   very important.   TRUE ❑ ❑ ❑ ❑ ❑ NOT TRUE

9. You admire farming/grower colleagues who are financially logical and don't let emotions colour their decisions.   TRUE ❏ ❏ ❏ ❏ ❏ NOT TRUE

10. You sometimes don't sleep at night worrying about decisions made.   TRUE ❏ ❏ ❏ ❏ ❏ NOT TRUE

11. You find investigating new farming/growing methods exhilarating and challenging.   TRUE ❏ ❏ ❏ ❏ ❏ NOT TRUE

12. You tend to write down options and calculate monetary consequences before deciding.   TRUE ❏ ❏ ❏ ❏ ❏ NOT TRUE

13. You tend to worry about what others think of your methods.   TRUE ❏ ❏ ❏ ❏ ❏ NOT TRUE

14. You are happy to make do with what materials you have to hand.   TRUE ❏ ❏ ❏ ❏ ❏ NOT TRUE

15. You find talking to others about growing ideas stimulates and excites you as well as increasing your enthusiasm for new ideas.   TRUE ❏ ❏ ❏ ❏ ❏ NOT TRUE

16. Having to make changes to well-established management systems and rules is a real pain.   TRUE ❏ ❏ ❏ ❏ ❏ NOT TRUE

17. You normally don't rest until the job is fully completed.   TRUE ❏ ❏ ❏ ❏ ❏ NOT TRUE

18. You normally enjoy being involved in farmer organizations.   TRUE ❏ ❏ ❏ ❏ ❏ NOT TRUE

19. You sometimes believe you are too much of a stickler for checking and double-checking that everything has been carried out satisfactorily.   TRUE ❏ ❏ ❏ ❏ ❏ NOT TRUE

20. When the pressure is on you sometimes become cross and short with others.   TRUE ❏ ❏ ❏ ❏ ❏ NOT TRUE

21. You generally choose conclusions from experience rather than from hunches when they are in conflict.   TRUE ❏ ❏ ❏ ❏ ❏ NOT TRUE

22. You are inclined to let employees/contractors do it their way.   TRUE ❏ ❏ ❏ ❏ ❏ NOT TRUE

23. You not only speak your mind and ask questions at farmer/grower meetings, but also enjoy the involvement.   TRUE ❏ ❏ ❏ ❏ ❏ NOT TRUE

24. It is very important to stick to management principles no matter what the pressure to do otherwise.   TRUE ❏ ❏ ❏ ❏ ❏ NOT TRUE

25. You are much happier if everything is planned well ahead of time.   TRUE ❏ ❏ ❏ ❏ ❏ NOT TRUE

Appendix 2B. Managerial Aptitude Test

Instructions

- Leave any question blank if you don't know the answer, or wish to bypass it.
- Answer by writing in the box the number of the correct answer where choices are given, **OR** the actual answer, **OR** tick the relevant box.

I. MEMORY

1. How many acres are in a hectare? ☐ acres

2. For sedimentary soils what is the desirable Olsen P test value? ☐

3. What is a desirable pH for good growth? ☐

4. What is the normal commission on stock/produce sales? ☐ %

5. What is *Trifolium repens*? ☐
   (1) white clover   (2) lucerne   (3) red clover   (4) wheat

6. In the RMA, what is a complying activity? ☐
   (1) One where community consultation approves.
   (2) One where there are no objectors.
   (3) One that is listed in the district plan.
   (4) One where the plans must meet the building standards.

7. How many instalments are there for provisional tax? ☐

8. The Occupational Health and Safety in Employment Act requires a producer to: ☐
   (1) Keep a register of accidents that harm an employee.
   (2) Report all illnesses that keep an employee in bed for more than 1 day.
   (3) Put a warning notice on all machines that could cause injury.
   (4) None of the above.

9. What is the current gift duty rate for total gifts less than US$27,000/annum? ☐ %

10. Which of the statements below MOST complies with the efforts to minimize worm resistance to drenches? ☐
    (1) Conduct faecal egg counts.   (2) Rotate drench types.
    (3) Rotate the mobs drenched.   (4) Minimize ectoparasites.

11. At the works, Standard superphosphate costs around? ☐
    (1) more than US$250/t.   (2) US$230–249/t.
    (3) US$210–229/t.   (4) less than US$210/t.

II. EXPERIENCE

1. Think back to a decision you made on feed management (e.g. to buy/sell a significant quantity of hay, to re-grass a paddock, to stop/start irrigation, to

use an area that was shut up for, perhaps hay, or perhaps winter use...) that, in hindsight, was very wrong.

> [ ]

What was this decision?

Describe the lessons learnt:

> 1.
> 2.
> 3.

Have you made this, or similar, mistake since, or previously?   Enter Y or N   [ ]

2. How do you work out the rules to follow when considering when to wean lambs?
   Enter the number of the description that is MOST appropriate.
   (Read ALL the options BEFORE answering)   [ ]

   (1) The locals and/or neighbours suggested the best rules.
   (2) I have discovered from past experience what is best.
   (3) I worked out the best rules based on my reading from magazines, books, and field-day handouts and the like.
   (4) An advisor/consultant told me the rules to follow.
   (5) Definitely a combination of most of the above.
   (6) Other.

3. Over the years, how much have you changed your management systems as a result of the hard lessons of less-than-hoped-for outcomes?
   Tick ONE box to describe the degree of change
   CHANGED A LOT ❏ ❏ ❏ ❏ ❏ NOT CHANGED

III. CREATIVITY

1. Assume your water supply for domestic and stock uses has come from rain-water and a reliable water race. This has been totally adequate. But, the water race system is to be closed down due to some resource consent problems. What do you think are the best two solutions that might be possible and should be investigated?

   (1) legal advice on the resource consent problem
       and consider the whole reason for the shutdown –
       can it be reversed?
   (2) Investigate wells and/or stream sources.   NO. OF BEST SOLUTION [ ]
   (3) Investigate a community water scheme.
   (4) Extend the rainwater collection area
       and storage capacity.
   (5) Put in more tanks and truck in water.   NO. OF 2ND BEST SOLUTION [ ]

2. What farming/horticultural problems would you recommend for research assuming quite limited funds? List, in priority order, the most important topics with respect to a good pay-off to the nation.

> 1.
> 2.
> 3.
> 4.

3. Assume you have purchased a new ploughable block next door to your back boundary that has an identical climate, and good soils. It also has a stream and water right for extensive irrigation. What are you going to do with the new block?

> 1.
> 2.
> 3.
> 4.

IV. GENERAL

1. What is out of place?
   (i) (1) Ryegrass (2) Phalaris (3) Alsike (4) Coxsfoot
   (5) Chewings fescue
   (ii) (1) Aberden Angus (2) Hereford (3) Charolais
   (4) Bos taurus (5) Jersey

2. List what you might call your management mistakes, if any, that have occurred over the last 12 months.

> 1.
> 2.
> 3.
> 4.

3. For my tax records and income tax return I do the following:
   Enter the number of the description that is MOST like your practise. (Read ALL the options BEFORE answering).

   (1) Prepare the tax return myself using my records.
   (2) Write up a cash book of all income/expenses, record the reference number of all source documents, and give the book and the files to my accountant.

(3) Use a computer to record all transactions and give the printout and/or disk to my accountant.
(4) Collect all invoices, statements, sale dockets, etc. and give them to my accountant.
(5) Other.

4. Which statement is Incorrect?
(1) Escort is for broom. (2) Tordon is for gorse.
(3) Versatil is for scrub. (4) Glyphosate is for grass.

5. A break in wool is caused by (put the number in the box)
(1) fungus. (2) a night of severe weather. (3) nutritional deficit.
(4) onset of longer days.

6. A knapsack is to herbicide as a drenchgun is to?
(Put the number of the answer in the box)
(1) anthelmintic (2) fungicide (3) sporadicide (4) innoculum.

7. Grandson is to grandfather as ram is to? (Put the number of the answer in the box)
(1) breed upgrade (2) grand dam (3) ancestors (4) progeny.

8. What is the next number in the lambing percentage series?
90 95 105 120…

9. You are told that the grass cultivar 'smart' is a selection of the cultivar 'slow'. Cultivar 'great' was bred from 'smart'. Thus, we must conclude 'smart' grows faster than 'slow'.
(T)rue or (F)alse?

10. Jack won some money in a growth rate competition organized by the drench suppliers. Jack spent it ALL in three competing stock and station companies. In the second store he spent US$100 more than half of what he did in the first, AND in the third US$100 more than half of the amount spent in the second. In the first store he bought, of course, US$500 of drench. How much did Jack win?

11. Jack has asked Tom to load the trailer as a prelude to a fencing job. Jack says they need 43 waratahs. Tom can carry five at a time. How many trips did Tom make?

12. If you rearrange the letters TPLSAE you would have the name of a
(1) sheep breed (2) clover (3) grass (4) fence component

V. SHAPES

1. John is working out how to subdivide a very large paddock(field) that has recently been successfully sprayed for broom. He is aware separating sunny and dark faces is important, and that stock drift uphill. So far his subdivision looks like:

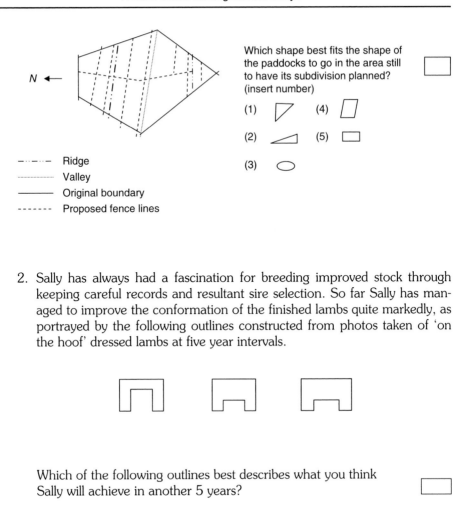

Ridge
Valley
Original boundary
Proposed fence lines

Which shape best fits the shape of the paddocks to go in the area still to have its subdivision planned? (insert number)

(1) (4)

(2) (5)

(3)

2. Sally has always had a fascination for breeding improved stock through keeping careful records and resultant sire selection. So far Sally has managed to improve the conformation of the finished lambs quite markedly, as portrayed by the following outlines constructed from photos taken of 'on the hoof' dressed lambs at five year intervals.

Which of the following outlines best describes what you think Sally will achieve in another 5 years?

(1)   (2)   (3)

3. Molly is a keen gardener and has put a lot of time into designing and planting her 'oasis' in the rather isolated place the homestead is located. The plant outline Molly wants between the edge of the front lawn and distant mountains is (Molly selected species accordingly, and won't prune):

Currently the form is:

Which of the forms below would you expect at the halfway stage?

(1)

(2)

(3)

VI. CALCULATIONS

1. You receive a call from your diesel delivery person wanting to know how much you need. You know the 1000l tank is about one-fourth full. The delivery person comes round about every 2 months, but sometimes it is as much as 3 months at most. While you don't do a lot of tractor work at this time of year you seem to use about 50l/week. How much should you order given you always try to never get the tank below one-fifth full (in litres)?

2. How many ewe lambs are you going to keep? Your flock, just past lambing, is currently 3000 ewes and in 2 years from now you want 3200 ewes and will NOT buy replacements. In the hogget flock (same number as last year) you have always had 2% deaths and cull about 15% on wool weights. The ewe flock is mixed age and, in the past, you culled 520 ewes per year. Ewe flock deaths average 3%.
Number of ewe lambs to keep?

   If the lambing percentage is 115% S to S, how many ewes should go to the ram for replacements?

3. Drench is on special – the price is the lowest you have seen it for this excellent drench. While you know it has a shelf-life of several years and resistance is not expected to be a problem, you reckon it is worth buying a 2-year supply. In the past you have concentrated on a 'clean pasture' policy through rotational grazing and haven't used a lot of drench. In fact, you have only drenched the ewes a couple of times per year, and the hoggets three times per year. The recommended dose is 2 ml/10 kg live weight. Which one of the following ranges covers how much you need to buy for your typical 3000 MA Romney ewes AND replacement flock? (Enter 1, 2, 3, or 4).
(1) <30l   (2) 30–60l   (3) 61–90l   (4) >90l

3 The Origins of Managerial Ability

Introduction

Chapter 2 introduced a range of factors likely to be involved in creating a farmer's managerial skill. It is known that a wide range of skill levels exists in any community. This variability can be used to relate these factors to outcomes achieved by using the data from any particular group of farmers. The importance of each factor may depend on the environment, but to assess this would require many sets of observations. This chapter, however, contains a discussion on the results of quantifying the relationship between the basic factors and outcomes for a large sample of all types in all environments, thus providing a generalized relationship. The farmers in the sample are relatively sophisticated with approximately a third having some form of formal tertiary education, and certainly all have at least 3 years secondary education. The farms are relatively large in terms of the number of people fully employed relative to worldwide averages, and in terms of the output per person employed. Many would involve an investment of at least US$1,000,000.

Determining the importance of each factor enables working out where effort should be directed in improving managerial skill, and indeed, in determining which factors can perhaps be left aside in such programmes. In assessing the success of a farm, it is necessary to counter the influence of resource quality to enable comparisons. A farm with high-quality soil, for example, would obviously have higher output per hectare relative to a poor-quality soil farm, even if the manager were rather poor.

Also relevant is the objective set held by the farmer. If a farmer has, say, an interest in maximizing leisure, it is not relevant to compare this farmer's profit with one with a solely maximum-profit objective. Thus, it is important to allow for the different objectives when assessing the origins of managerial ability. A farmer maximizing leisure might well be a very efficient manager, but when compared with the profit per hectare of other farmers he may well rank poorly.

Thus, ways of measuring the objectives is introduced as a prelude to its inclusion in a quantified model.

Similarly, a part of managerial skill could well be a farmer's attitude to how much control he believes he has over outcomes. Some farmers believe, for example, that they have little influence due to the variability and impact of important factors such as the weather and markets. A measure of this control belief is called a farmer's 'locus of control' (LOC). A test for measuring this belief is introduced and commented on. Furthermore, an assessment of how it relates to measures of success is introduced for a sample of farmers.

Objectives

As noted, to enable comparisons it is essential to quantify a farmer's objectives. Furthermore, it is possible that a farmer's objectives will in their own right influence managerial ability. A farmer who is quietly content to enjoy primary production for its intrinsic values might well be content to be an 'average' manager and spend little time improving his skills.

The results of quantifying a set of objectives through a formal questionnaire can be used in conjunction with the casual observations that anyone dealing with a farmer will inevitably make. The outcome should be a good assessment of the objectives. Observing the actual decisions made by a farmer will indicate the form the farmer's objectives take. Furthermore, quantifying the objectives involves writing them down, and this is often a catalyst to reviewing and more carefully defining what the farmer is striving to achieve. This process in itself can be beneficial. This is particularly the case where the farmer has not formalized a list of objectives, simply preferring to follow their feelings on each issue. This is quite common.

A complicating factor is where there are many owners of a farm, or where a family is involved, each member of which may well have issues with the appropriate objectives to follow. Clearly, a simple case of just one owner living on the farm makes deciding on the objectives very simple. Where this is not the case, it is useful for each interested party to write out their objectives and use the lists as a basis for a round table discussion in a move to obtain a consensus.

There are many formal question sets available in the literature that have been used to assess objectives. Listed below is one example in which the farmer is asked to rate the truthfulness associated with a series of statements. It will be noted that all the major aspects that might be of interest to a farmer and the family are included.

GOALS AND AIMS

Tick ONE box that best records your degree of belief in the statements.

1. It is very important to pass on the property
   to family members.   TRUE ❏ ❏ ❏ ❏ ❏ NOT TRUE

2. It is important to earn the respect of
   farmers/growers in the local community.   TRUE ❏ ❏ ❏ ❏ ❏ NOT TRUE

3. Making a comfortable living is important.   TRUE ❏ ❏ ❏ ❏ ❏ NOT TRUE

4. It is very necessary to keep debt as
   low as possible.   TRUE ❏ ❏ ❏ ❏ ❏ NOT TRUE

5. It is essential to plan for reasonable holidays
   and plenty of leisure time.   TRUE ❏ ❏ ❏ ❏ ❏ NOT TRUE

6. Attending field days and farmer/grower
   meetings is vital.   TRUE ❏ ❏ ❏ ❏ ❏ NOT TRUE

7. It is very important to reduce risk using
   techniques like diversification, farming
   conservatively, keeping cash reserves, etc.   TRUE ❏ ❏ ❏ ❏ ❏ NOT TRUE

8. Developing facilities and systems that
   give good working conditions is crucial.   TRUE ❏ ❏ ❏ ❏ ❏ NOT TRUE

9. It is very important to ensure employees
   enjoy their jobs.   TRUE ❏ ❏ ❏ ❏ ❏ NOT TRUE

10. Doing jobs that I enjoy is a very important
    part of the operation.   TRUE ❏ ❏ ❏ ❏ ❏ NOT TRUE

11. Minimizing pollution is very important.   TRUE ❏ ❏ ❏ ❏ ❏ NOT TRUE

12. I enjoy experimenting with new products
    and production systems.   TRUE ❏ ❏ ❏ ❏ ❏ NOT TRUE

13. Proper retirement planning is a major
    consideration.   TRUE ❏ ❏ ❏ ❏ ❏ NOT TRUE

14. You must always be striving to increase
    the total value of assets.   TRUE ❏ ❏ ❏ ❏ ❏ NOT TRUE

15. Constantly expanding the size of the
    business is absolutely necessary.   TRUE ❏ ❏ ❏ ❏ ❏ NOT TRUE

16. Aiming for maximum sustainable net
    cash returns is very important.   TRUE ❏ ❏ ❏ ❏ ❏ NOT TRUE

17. Maintaining a presence in local community
    activities is important.   TRUE ❏ ❏ ❏ ❏ ❏ NOT TRUE

18. It is very important to improve the
    condition of the property
    (fertility, facilities, etc.).   TRUE ❏ ❏ ❏ ❏ ❏ NOT TRUE

19. Giving assets to the children so they can
    pay for education and/or set up businesses
    is very important.   TRUE ❏ ❏ ❏ ❏ ❏ NOT TRUE

20. While I don't particularly enjoy farming,
    I carry on as I don't have a background
    that allows shifting into another occupation.   TRUE ❏ ❏ ❏ ❏ ❏ NOT TRUE

From the results of dairy farmer Hank's answers to the questionnaire it was clear he was very interested in the wealth factor as an objective, but he also believed 'family and community' was reasonably important in guiding his decisions. For 'fun and leisure', Hank had an average score as did the 'work enjoyment' factor. His responses reinforced his answers to the risk attitude test with the 'minimize risk' objective featuring quite strongly. It is interesting, and reassuring, to compare his objectives with his actions. Currently Hank is developing an additional farm purchased not long ago, thus helping satisfy his wealth objective. He works long hours, but does get involved in community affairs, so again complying with the questionnaire answers.

Margrave responded similarly over the wealth factor. Of course, you would expect excellent and ambitious managers to have this objective. But there again, some excellent managers might well be more interested in family, work enjoyment and community and therefore they must be judged on what their objectives are, not an observer's assumptions on profit maximization. For 'work enjoyment', Margrave had a score in the middle of the road, but he found 'family and community' not particularly important. This probably reflected in part that his spouse was in full-time work in the latter years. 'Fun and leisure' was similarly not very important, reflecting his passion for agriculture in contrast to other activities. Finally, his 'minimize risk' score showed this was not a major concern when making decisions. This also reflects his risk attitude.

Note that a further more detailed examination of objectives, their origins and the relationship between objectives and families is provided in Chapter 7.

Locus of Control

For farmers who have little confidence in their ability to control outcomes, the inference is that risky events, such as the weather, the state of the market, and the international exchange rate, are more important in the success of their farm than the decisions made. For exactly the same environment, however, some farmers will believe they have a reasonable degree of control over outcomes. Of course, just where the truth lies will depend on the environment and situation each farmer finds himself in. Certainly, in situations where the weather, markets and all the other uncertain areas are extremely variable and prediction is not possible, and where the number of alternate products and production methods is minimal, the farmer will have less control compared with an environment more stable and predictable with a wide range of choices.

Despite the environmental situation, it is useful to know just how much control a farmer believes is possible. This belief will impact on efficiency and skill. The reasons a farmer holds any one belief will depend on past experiences and the ability to relate decision with outcome. Comparisons with neighbouring farms also help formulate a conclusion. In addition, the lessons learnt while acquiring managerial skill, and the influence of other managers and parents will have influenced the attitude.

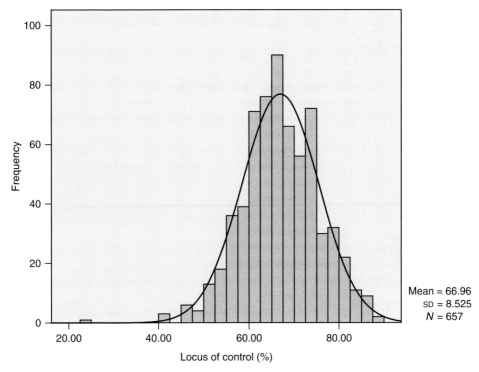

Fig. 3.1. Distribution of the respondents' 'locus of control'.

A test for assessing the degree of control belief is called the 'locus of control' (LOC). A farmer is classified as an 'internal' if he believes he has considerable control in contrast to a farmer who has an 'external' LOC. Of course, the situation is not as black and white as this, with a continuum existing from one end of the spectrum to the other. A wide range of question sets to assess the LOC have been developed for a wide range of interest areas (e.g. control of a person's health, work place safety, motor accident control, and so on). Listed below is a question set designed for agriculture that is based on one of the first general tests developed, but with the terminology and situations changed to relate to farmers.

QUESTION SET TO ASSESS A FARMER'S LOCUS OF CONTROL

For each of the following statements indicate how true it is. Each statement has five boxes beside it – tick only the ONE that best describes your degree of belief in the statement.

1. So far I have managed to largely achieve my goals.   TRUE ❏ ❏ ❏ ❏ ❏ NOT TRUE

2. I never try anything that might not work.   TRUE ❏ ❏ ❏ ❏ ❏ NOT TRUE

3. I'm using exactly the same production methods that I have used for many years as they have stood the test of time.   TRUE ❏ ❏ ❏ ❏ ❏ NOT TRUE

4. It's no use being stubborn about a job or management approach that doesn't initially work.   TRUE ❏ ❏ ❏ ❏ ❏ NOT TRUE

5. I reckon 'good luck' doesn't exist – 'luck' is really good management, and 'bad luck' poor management.   TRUE ❏ ❏ ❏ ❏ ❏ NOT TRUE

6. It is safer not to rely on others to get the job done well and on time.   TRUE ❏ ❏ ❏ ❏ ❏ NOT TRUE

7. I'm able to get others to do the jobs my way.   TRUE ❏ ❏ ❏ ❏ ❏ NOT TRUE

8. Too often I end up having to run my property to suit others' demands.   TRUE ❏ ❏ ❏ ❏ ❏ NOT TRUE

9. While being a good manager involves some training, experience and reading, management skill is mainly determined by your genes.   TRUE ❏ ❏ ❏ ❏ ❏ NOT TRUE

10. You can work hard at creating good relationships between neighbouring managers, but often your efforts fall on deaf ears as people are commonly uncooperative and self-interested.   TRUE ❏ ❏ ❏ ❏ ❏ NOT TRUE

11. I find most employees work hard and finish the tasks set very adequately after a bit of training where necessary.   TRUE ❏ ❏ ❏ ❏ ❏ NOT TRUE

12. The years when the property has shown poor production and profit have been due to circumstances totally out of my control.   TRUE ❏ ❏ ❏ ❏ ❏ NOT TRUE

13. In local body affairs it's easy for a hard-working and dedicated individual to have an impact in getting changes for the better.   TRUE ❏ ❏ ❏ ❏ ❏ NOT TRUE

14. Often I get frustrated as circumstances beyond my control impede thesmooth progress of my management plans and decisions.   TRUE ❏ ❏ ❏ ❏ ❏ NOT TRUE

15. Some people seem to be just lucky and everything works out for them, but it hasn't happened to me much.   TRUE ❏ ❏ ❏ ❏ ❏ NOT TRUE

16. I tend to carefully plan ahead to ensure my goals are achieved, andoften do budgets and commit my ideas to paper.   TRUE ❏ ❏ ❏ ❏ ❏ NOT TRUE

17. I seldom change my management and
    production systems unless I'm doubly sure
    the change will be positive. So much
    depends on chance.   TRUE ❏ ❏ ❏ ❏ ❏ NOT TRUE

18. When things go wrong it is so often due to
    events beyond my control – the weather
    ruins the hay, the wool auction I choose has
    a sudden price dip, etc.   TRUE ❏ ❏ ❏ ❏ ❏ NOT TRUE

19. When I know I'm right I can be very
    determined and can make things happen.   TRUE ❏ ❏ ❏ ❏ ❏ NOT TRUE

In developing an overall score for a respondent's belief, the score on each question is added after the score is reversed for the statements where 'true' means low control so a high total score means a strong belief in controlling outcomes. The scores can then be converted to a percentage figure, so a score of 100 indicates a very strong control belief.

As an example of the kind of control scores that are typical over a wide range of farming types in a temperate climate (New Zealand), the distribution data obtained from a sample survey is given in Fig. 3.2 together with the normal distribution.

The average score was 67%, and the standard deviation 8.5.

Clearly a wide range of control beliefs existed in the community, though the majority falls within the 55–80% band.

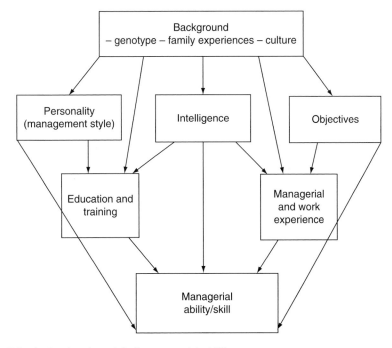

Fig. 3.2. A structural model of managerial ability.

As there are similarities in some of the control statements, it is useful to look at the correlations between statements to explore whether there are groupings that summarize the basic components of the farmer's control belief. This analysis suggested that seven factors underlie the beliefs. These can be called:

*   beyond control;
*   conservative traditionalist;
*   determined despite bad luck;
*   flexible achiever;
*   gene-based traditionalist;
*   careful and determined planner.

Producers with a high proportion of factor one have little control belief and accept that outcomes are largely luck, whereas people with a high proportion of factor two believe you make your own luck through using tried and tested methods. Factor three relates to being able to ride through bad luck, whereas factor four involves a belief that it is possible to achieve if you are flexible. In complete contrast, factor five embodies an acceptance that your genes determine ability, but despite this good achievement is possible. Finally, factor six is the belief that careful planning does work and will have beneficial pay-offs.

Any one manager will have a mix of these factors leading to an overall control belief. If a farmer's LOC is low, or excessively high, thought should be given to whether efforts to change it are appropriate. This is achieved by discussing the farmer's beliefs with examples of what is possible. Use of case examples to show what is in fact possible may well assist this process of creating a more realistic attitude. An excessively high LOC may lead to excessive risk taking where techniques to reduce risk are not deployed. An example is diversification.

Research has indicated that age does correlate with the LOC with youth tending to overestimate just how much control is possible. As in most things, appropriate experience, and an ability to learn from this experience, is a very valuable part of becoming a skilled manager as will be shown and discussed in following sections.

Margrave, who operates in a very variable climate, has a LOC of 57%, which is somewhat less than the population average of 67%. Yet he is certainly positive in his view of management with constant change to improve the outcomes. Margrave comments that 'with high risk levels you need to be creative' and 'you can mitigate many things if you think out of the square'. Hank, perhaps due to the high level of irrigation, has a score of 73%, expressing the control he certainly does have. This is also reflected in his purchase of extra land with huge investments in infrastructure.

As for the future, Hank asks 'How can I exert more control?'. Hank believes there will be increasing volatility in prices and markets as the years roll by due to the rapid communication systems that now abound. In the past, the happenings in a market 20,000 km away was hardly noticed, and only affected prices in the next season. However, now manufacturing and supply corporations are watching the daily movements in key markets and altering farmer payouts accordingly. The same sort of thing is happening in input markets. A good example is the fuel market where over 2 weeks the price at the pump might rise two times, and then suddenly decline only to rise again after 10 days. One wonders what the point of all the ups and downs is, but it certainly makes planning rather difficult. Therefore, Hank concludes, the farmer must get used to these ups and downs for they will continue for many years. As farmers have little control over the world markets, they must be downgrading their views on the degree of control they have.

Relating LOC to Managerial Factors

There have been a number of quantitative studies relating LOC to various aspects of management. They are not totally conclusive on the importance of a farmer's LOC.

Overall, it does appear that the LOC is more of an indication of a farmer's propensity to be involved in innovations than their actual managerial ability. Furthermore, as you would expect, management style does impact on a farmer's LOC, so it could be possible to infer a farmer's LOC from an understanding of their personality.

The research results show that farmers tending towards 'internality' are more likely to be adventurous in considering innovations, and tend to experience improved financial results relative to 'externals'. Similarly, internals are more likely to seek out and use extension activities and, consequently, learn about possible innovations.

It is also suspected that internals will have created more wealth than their external counterparts. This is a consequence of their innovativeness (Kaine et al., 2004).

For the New Zealand data, the farmer's LOC was related to their self-assessed managerial ability. The farmers were asked to score themselves on a ten-point skill scale relative to their colleagues. It was found that:

$$\text{Skill} = 3.3 + 0.08A - 0.5E + 0.22I + 0.01L$$

where L is the LOC percentage score, A is age in years, E is the level of education on a five-point scale (1 = primary education only through to 5 = three or more years tertiary education) and I is self-assessed intelligence on a four-point scale (1 = below average intelligence through to 4 = highly intelligent). This equation explained 32% of the variance in the skill level, leaving another 68% unexplained. It is interesting to note the negative coefficient on education in contrast to the impact of self-assessed intelligence, and age. LOC does have an impact so that, for example, if the LOC percentage changes by 20% the skill rating changes by 0.2 on the ten-point scale. However, this is not a major impact.

Despite this low impact of LOC on self-assessed skill, a farmer's control attitude undoubtedly does impact on choices in that an 'external' is likely to choose alternatives where the farmer perceives greater control is likely. Thus, for example, where the economics of irrigation is marginal, an 'external' is likely to proceed with an investment as the irrigation will give more control and surety.

It is also interesting to note that the correlations between the LOC and personality measures. In a study of New Zealand farmers of all farm types there was a:

- 26% (32% in another survey) correlation between the LOC and the anxiety personality factor;
- 44% correlation with conscientiousness; and
- 12% (21%) correlation with extraversion.

The other two traits were not significantly correlated other than a 9% (35%) correlation with openness. You would expect these relationships for an 'open' person can be described as 'original, daring and liberal' and is, therefore, likely to believe in having control as would a person who is an extrovert (sociable, spontaneous), though the degree of control is not as great. On the other hand, an anxious person would have a lack of a sense of control, and this is the case with the 26% (32%) correlation in the opposite direction to the others. Given these quite high levels of correlation it could be suggested that a study of farmers' personalities would tell you almost as much as a separate LOC test. To further examine these factors Chapter 7 contains a discussion on variables that might be regarded as giving rise to a farmer's LOC.

Modelling Managerial Ability

Factors giving rise to ability

Understanding the factors that give rise to a farmer's managerial ability leads to being able to consider how ability might be improved. Furthermore, having data on the quantitative impact of each factor enables concentrating on the most important aspects of ability.

It has been pointed out that, in particular, management style and intelligence are likely to be important determinants of ability. Equally, you would expect experience to be important, particularly where the farmer has the attributes to gain benefit from the experiences. Unfortunately, it is difficult to measure experience as there are no simple quantitative measures. Age is certainly a factor, as it is more likely for an older farmer to have been exposed to potentially beneficial situations. It is easy and important to include age as an explanatory variable.

Also likely to be relevant will be a farmer's formal education, both in terms of the number of years, but also the educational success achieved. There is strong evidence from urban business situations that employees' incomes are highly correlated with education. It is also interesting to note that a major determiner of school attainment is family background. With respect to society's investment in schooling, the return for each additional year of education is around 9% (Ashenfelter and Rouse, 1998). This figure, or very similar, is quite common in several studies. You would expect that, similarly, the return to education through managerial ability will be positive.

The other factor likely to impact on ability is the farmer's objectives, or where they are taken into account, the amalgam of the family members' objectives. Thus, the earlier discussion on assessing objectives is due to their likely impact on farming success, both in the sense of measuring the right outcomes when assessing success, and for their direct impact on efficiency.

In summary, the factors likely to determine a manager's managerial ability are his:

- management style (personality);
- intelligence;
- education;
- experience; and
- objectives.

Also possibly relevant are both the family background, and the prevailing factors that might influence skill that are dominant in the society within which the farmer grew up and operates. Of course, the farmer's family background helps determine his educational experience (it was noted above that family background is a major influence on educational success), but also practices such as involving the children in discussions on farm decisions. This involvement could impact on the development of managerial skill, as could the involvement in the physical work. Similarly, the dominant culture of the time could influence managerial development so, for example, if it was accepted in the community that children were included in any extension meetings and field days, the experiences could have positive impacts. Another example might be whether children were sent off to boarding schools, in contrast to using local schools, and whether they should go to tertiary education, or learn on the job.

Understanding the importance of all these factors will assist in setting up programmes to ensure young managers are as skilled as humanly possible.

Modelling results

It is easy to draw a diagram of the factors that might be important. It should show which factors influence other factors that in turn determine the level of managerial ability. It is another matter, however, to work out just how important each factor is. A diagram representing the factors might look like Fig. 3.2.

What is required is a knowledge of how important each arrow is in determining each of the factors in the diagram. The direction of the arrows suggests that the background factors impact on the farmer's managerial style, intelligence and objectives. These in turn impact on the education and experiences. Finally, in addition to these factors, management style, intelligence, objectives, education and experience all also directly influence the farmer's managerial ability.

To put figures on the arrows' 'importance' requires a technique called 'structural equation modelling'. This works out the regression parameters giving the best fit of the data to the model. If you like, a whole series of linear regressions is carried out to find out the equation which best fits the prediction of the variable in each box. Fortunately, computer packages are available to work all this out, so it is simply a matter of defining the form of the model, and then providing the package with the data available for the sample of farmers.

In this case, data for most variables (as described in each box) was available for the sample of New Zealand farmers described previously and was used to hypothesize, and to calculate the parameters for, a structural equation model of ability. This is shown in Fig. 3.3. The ellipses represent variables that are not actually observed, while the boxes represent the observed variables – in this case for each farmer in the sample. Thus, information was available on the farmers' parents and forebears as well as their grades for their last year of formal education, highest level of formal education, age, gender, as well as their self-rated managerial ability (or skill). In addition, their locus of control rating was available, as well as the average of their last 5 years' profit increase (sometimes negative), and total asset value increase, and their level of physical output productivity relative to their peers. Given the belief that experience was important, data on a whole range of experience factors was gathered and gave rise to the length of experience (time–experience) variable, and to the value of different types of experience (learn–experience) variable. Finally, information on the farmer's management style and objectives, as described previously, was also available.

The model hypothesized that the unobserved 'true ability' is dependent on a farmer's management style, the sum of the experience factors (management experience) and the managers' 'true intelligence'. This latter variable was

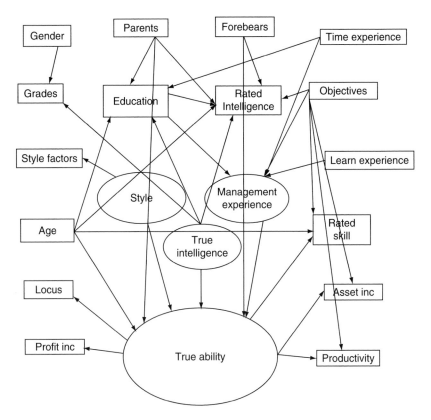

Fig. 3.3. A structural equation model of managerial ability.

inferred from the farmer's grades, level of education and self-rated intelligence. Education, in turn, depends on the farmer's parents, age (younger farmers had greater opportunities) and experience time. By examining the direction of the arrows, the sources of the basic variables can be inferred. It will be seen that the farmer's objectives are assumed to influence a number of variables such as the farmer's self-rated ability (skill). Thus, an allowance is made for the nature of the farmer in assessing his self ratings.

The results of the analysis provided standardized regression coefficients for the influence of each impact represented by an arrow. The critical results are the experience→ability arrow with its coefficient of 0.971 relative to the influence of 0.232 for style and 0.109 for true intelligence (the advantage of giving the standardized coefficients is that they provide direct comparability having allowed for the different units of measure used). Thus, experience is approximately four times as important as style in determining managerial ability, and intelligence only a ninth as important. These are important conclusions indicating very clearly the origins of managerial ability. It is worth noting that the second survey data gave similar conclusions. It is important to comment this does not mean intelligence is not important for, clearly, to learn the lessons provided by experience a certain degree of education and intelligence is important.

To further reinforce information on the important drivers of managerial ability, the farmers in the survey giving rise to the data above were divided into two groups based on their true ability. By comparing the groups, clues as to the most important attributes are provided. The arbitrary division was based on the 70% ability line after converting the farmer's scores to a percentage, the distribution of which is given in Fig. 3.4.

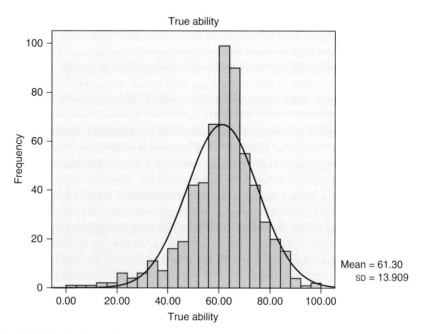

Fig. 3.4. Distribution of farmers' true managerial ability.

14. Pasture

Divide your grazing pasture into four groups* based on their dry matter cover (make your best guesstimate)

(i) Worst _____   has of _____ kg DM/ha
(ii) Basic _____   has of _____ kg DM/ha
(iii) Reasonable _____ has of _____ kg DM/ha
(iv) Best _____   has of _____ kg DM/ha
* include any areas shut up

15. Lucerne (alfalfa) – leave blank if none.

Divide your lucerne (alfalfa) into four groups* based on their dry matter cover (make your best guesstimate)

(i) Worst _____   has of _____ kg DM/ha
(ii) Basic _____   has of _____ kg DM/ha
(iii) Reasonable _____ has of _____ kg DM/ha
(iv) Best _____   has of _____ kg DM/ha
* include any areas shut up

16. Feed crops – leave blank if none.

Divide your feed crops into four groups* based on their dry matter cover (make your best guesstimate)

(i) Worst _____   has of _____ kg DM/ha
(ii) Reasonable _____ has of _____ kg DM/ha
(iii) Best _____   has of _____ kg DM/ha
* include any areas shut up

17. Cash crops – leave blank if none.

Divide your cash crops into two groups.

(i) Looking like average or better yield   has
(ii) Looking like less than average yield   has

18. TOTAL PRODUCTIVE AREA IS (add up above)   has. Is this total correct?

19. What is your most productive grass variety/cultivar?

20. What is your most productive clover variety/cultivar?

21. Hay reserves.

Divide any hay/baylage into three quality levels and give:

No. of bales (average _____ kg/bale) of   Dominant species
                                    high-quality feed
No. of bales (average _____ kg/bale) of   „
                                    medium-quality feed
No. of bales (average _____ kg/bale) of   „
                                    low-quality feed
How many bales are not adequately covered?

22. Silage reserves (leave blank if none)

Divide any silage into three quality levels and give:

No. of tonnes of high-quality feed   Dominant species
No. of tonnes of medium-quality feed   „
No. of tonnes of low-quality feed   „

23. Straw reserves (leave blank if none)

How many bales (average _____ kg/bale) of straw do you hold as a feed source?

24. Concentrate/grain (leave blank if none)

How many tonnes of concentrate/grain are held as a feed reserve?
Tonnes   Dominant type

25. FENCES (leave blank if not applicable)

Average size of paddocks/fields   has.   Biggest   has
                                    Smallest   has
                                    Most common   has
Distance of good-quality fences   km
Distance of average-quality fences   km
Distance of poor-quality fences   km
Main type of fence _____ (describe, e.g. posts & netting)

No. of paddocks/fields with good shelter
                          average shelter
                          poor shelter
No. of paddocks/fields with good water
                          average water
                          poor water
Overall reliability of water supply (good, OK, poor)
Current average 'fullness' of dams/tanks   %

26. BUILDINGS

Woolshed – no. of stands   no. of sheep under cover
           Condition (poor, OK, excellent)
           Area (m²)?
Sheds:
     Type   Area (m²)   Condition (poor, OK, excellent)
     Type   Area (m²)   Condition (poor, OK, excellent)
     Type   Area (m²)   Condition (poor, OK, excellent)
Hay barns:
     No.   Total capacity   bales of kg/bale
     Average condition (poor, ok, excellent)
Yards/enclosures:
     sheep – capacity   sheep   Condition (poor, OK, excellent)
     Cattle- capacity   head   Condition (poor, OK, excellent)

27. MACHINERY (leave blank if not applicable)

      Tractor one wattage   age   years Condition (good, average, poor)
      Tractor two wattage   age   years Condition (good, average, poor)
      Tractor three wattage   age   years Condition (good, average, poor)
         Average capacity of truck/s   tonnes
         Header size   metres cut?
         Baler size(kg/bale)   age   years Condition (good, average, poor)
         No. of four-wheeler bikes   Condition (good, average, poor)
         No. of items of cultivation machinery
         Value of the irrigation plant (if any)

28. PHYSICAL

      Area flat land   has   Soil type?
      Current available soil moisture level _____   %
      Area rolling but cultivable   has   Soil type?
      Current available soil moisture level _____   %
      Area hill but NOT cultivable   has   Soil type?
      Current available soil moisture level _____   %
      What area is irrigable?   has
      What area do you irrigate?   has   How much water per hectare?
      Rainfall:
      Average (mm) spring   summer   autumn (fall)   winter
      Highest might expect spring   summer   autumn (fall)   winter
      Lowest might expect spring   summer   autumn (fall)   winter
      How much rain (mm) so far this year?

29. PASTURE AND LUCERNE (ALFALFA) PRODUCTION

On average, what do you expect from your pasture/lucerne (alfalfa)?

                                            Pasture   Lucerne (alfalfa)
      Good season   kg DM/year
      Typical season   kg DM/year
      Poor season   kg DM/year
      Highest growth time (month?)   kg DM/day
      Lowest growth time (month?)   kg DM/day

      What is your current stocking rate?   S.U./ha
      What do you think is the maximum possible?   S.U./ha

A 'S.U.' (stock unit) is the feed necessary to support a 50 kg ewe.

30. 'ON-HAND' MATERIAL

      How much fertilizer of all kinds is currently held?   tonnes
      How much grass/clover seed is on hand?   kilograms
      How much spray material (all kinds) on hand?   litres
      How much drench of various kinds is on hand?   litres
      How many kg of wool is 'on hand' waiting sale?   kilograms
      How much grain/seed in silos is waiting sale?   tonnes

31. FINANCIAL

What is the net balance of all your trading   Credit/Deficit
bank balances?
What money is currently owed from
invoices/statements?
What money is currently owed to you? (Credits held)
Mortgages and Loans – what is the current outstanding
debt for the farm as a whole?

For each loan give amount, amount outstanding, principal repayment/
year, interest/year, and total payment per year (in US$)

Loan one
Loan two
Loan three
Loan four
Loan five

* Note – if you don't know the split between principal and interest just
put in the total payment/year.

32. ASSET VALUE – for the whole farm unit:

What is the total current market value of your
land and buildings?   US$_____
What is the total current market value of your
machinery & plant?   US$_____
What is the total current market value of your sheep?   US$_____
What is the total current market value of your cattle?   US$_____
What is the total current market value of all other assets?   US$_____

33. NET WORTH

Your current net worth is (sum of assets minus sum of
loans and debts outstanding)   US$_____

If any, how much more do you think you could borrow
if necessary?   US$_____
What proportion of the amount, if any, would you be
happy to borrow?   (%)

How many of the applicable questions could the farmer answer with
certainty?
And how many was the farmer only partially sure about?
And for just how many did the farmer take what might be called a guess?

If the farmer does not know with reasonable accuracy just what situation
the farm and the assets or debts are in, it is not possible to make the best
plans. Good management starts from observation and a good knowledge of
the current state of the farm.

4   Decision Processes and Goals

Introduction

While each farmer will have his own particular way of making decisions, there are some common themes. Understanding the possibilities assists in following why a farmer makes a specific decision, and in discussing how the decision process might be improved.

Generally, a farmer observes factors from the surrounding environment, processes this information, and comes to a conclusion about what actions to take, one of which might be to do nothing just yet. Thus the following sequence occurs on a daily basis, irrespective of whether the decision horizon is a few hours or several years.

$$\text{Stimuli} \rightarrow \text{sensory systems} \rightarrow \text{short-term memory} \rightarrow \text{long-term memory}$$
$$\text{conscious processing} \leftarrow \text{retrieval}$$
$$\text{subconscious processing} \leftarrow \text{retrieval}$$

Many of the stimuli being received are acted on automatically, so, for example, we avoid stepping in a hole in the road without realizing we have sorted this little problem. Other stimuli require conscious thought (a bad storm is forecast) with the information being processed with a conclusion to, say, take action (shift the lambing mob of ewes into a sheltered area). Some stimuli get considered and then sent to long-term memory as being important. It can then be retrieved when required. For example, perhaps a newspaper reports that the normal spring weather in a country producing a significant quantity of wheat has not occurred. This information might be stored as something to act on later if further evidence comes to hand that the wheat production is, worldwide, lower than normal.

When talking about managerial ability, we are concerned with the success of this decision process, particularly the observation systems, and both the conscious and subconscious processing systems. If all these operations work appropriately, we end up with good management. Also relevant is the memory storage and retrieval system. Clearly appreciating the existence of all these systems helps understand good management, and what might be done to improve management.

Besides the decision processes used, the goals and objectives are important for they refine the details of the processes used. If the objective is to simply maximize profit, then the processes used in deciding on appropriate actions should be different to where not only cash income is important, but also, say, the risk level experienced in earning the income. Thus, this chapter contains a discussion on both the processes that might be used in the processing stages, and their relationship to a farmer's goals.

Any decision must be stimulated by the manager recognizing that some sort of problem exists. This word 'problem' is used in its general sense to define any situation where some kind of thought and analysis is required. Thus, for example, when a new herbicide is available, there is a problem in the sense that a decision is required on whether it should be used in preference to the current chemical. Also, of course, problems in the normal sense of the word require decisions so, for example, when the crop is not growing as expected, a decision on whether additional fertilizer, or perhaps irrigation, should be applied is necessary. Whatever the case, successful observation, analysis and decision making is required to direct the farm on to the optimal path.

The simplest decision process, which is not uncommon, is the 'do as the others do' approach. Quite often in any farming community some participants are regarded as being the leading farmers, so the others watch what they are doing (by direct observation, by talking to the leading farmer(s) or perhaps talking to other members of the community who know what is happening). Management then consists of simply 'following the leader'. Whether this succeeds obviously depends on how good the leaders are, whether their situations are comparable, whether the implementation abilities of the followers are suitable for the system adopted and whether their objectives are the same. Having all these attributes occurring at the same time is not common.

The other extreme is where a farmer diligently researches every decision situation and has the skills and knowledge necessary to successfully carry out a thorough and logically correct analysis, and then the ability to follow it through. In between these extremes are a range of processes which will be outlined. Of course, in reality there are an infinite range of approaches taken by farmers with a continuum occurring between the extremes. Only a sample set can be summarized.

What follows is a synopsis of both the main processes, and how they interact with the objectives. It is assumed that a problem has been identified, so it is then a matter of sorting out its structure, gathering data, analysis, decision, action, feedback and starting the whole process again.

Decision Processes

Linear approach

One theory is the standard rational approach of carrying out a number of steps in a simple linear order. A common list of tasks is:

- information gathering;
- data evaluation;
- problem structuring;
- hypothesis generation;
- preference specifications;
- action selection;
- decision evaluation.

There are many variations on the tasks and ordering that appear in the literature, but suffice to note that the concept is that of a formal set of operations carried out in sequence until a decision is made, which is then implemented. The idea is that a hypothesis is only generated once the data has been collected and examined, so it is assumed that this hypothesis contains the list of likely alternative conclusions. These are then ranked on the basis of their contribution to the objectives. Others might suggest that a hypothesis, or idea, is proposed right at the beginning for it is only once some structure is sorted out that you know what data to collect. In fact, it is likely very few people act entirely rationally as is suggested with this linear process.

It was noted earlier that short-term memory has limited capacity to hold material. Thus, in working through the linear process it is necessary for subsection conclusions to be sent off to long-term memory, so that the process proceeds in sections with all the information being finally brought together. Success in this kind of analysis improves with experience in that practice hones a knowledge of how to break up the analyses into sections. There is evidence to suggest that with complex problems, and many are of this form in agriculture given all the disciplines involved, do get broken into simplified chunks. The ability to sort out just what are the key issues is clearly important for success. For example, decisions about the use of phosphate fertilizer involves a knowledge of:

- soil chemistry;
- distribution patterns;
- plant requirements;
- soil organic matter;
- water patterns and soil-holding capacity; and eventually
- response rates and plant nutrient requirements.

No doubt there are more aspects than these that are also important. However, it is possible to simplify the whole analysis if it is realized that soil colloids bind the phosphorus which not only restricts leaching, but also provides a buffer of the element that is supplied over many years. Such simplifications that still allow a correct analysis and decision to occur provide help in actually reaching a rational decision for there is a clear limit to the brain power of all decision makers.

Indeed, a feature of 'experts' is that they are able to sort out the critical parameters and thus discard all other information that might confuse the issue.

Decision rules

No matter what the approach, linear or otherwise, the farmer's knowledge of the appropriate decision rules is crucial. A study of production economics allows working out what rules to use in making an optimal decision. Research has shown that most production follows a sigmoid curve in that as more inputs are added, production increases at an increasing rate, but with even more input, this increase declines to provide decreasing returns. Eventually the shape of the curve describing output to input starts to bend backwards indicating output is actually falling as more and more input is added. For example, if large quantities of nitrogen fertilizer are applied to a crop, eventually it becomes too lush and falls over, possibly with a disease infestation. The excess nitrogen might even be toxic.

Under this classical production function, as the relationship between input and output is called, where the output price is sufficient to cover the basic costs so some level of production is economic, it can be proved that the decision rule for maximizing profit is to:

- move along the production function until the marginal cost of the last unit of input added is just equal to the marginal return from the increased output from this last unit of input (Decision rule 1).

Due to the diminishing returns, going past this point means the increased cost of the extra fertilizer added is not covered by the resultant increase in production. For a full explanation, any standard text on production economics can be referred to.

This rule tells 'how much', per production unit (e.g. hectare of land), to produce. There is also the question of:

- second, what inputs to use; and
- third, what products to produce.

As you would expect from the description above, production economics 'derives' the decision rules that guide these decisions for optimizing an objective, provided it can be measured. Assuming an objective of maximum cash profit, inputs should be substituted for each other to the point where the:

- marginal rate of substitution between them multiplied by their cost is just equal (Decision rule 2).

The marginal rate of substitution is the amount of, say, a fertilizer that is required to keep output at the same level as it is substituted by another input, say irrigation. If, for example, 1 kg of nitrogen fertilizer could be substituted by 4 mm of irrigation water so that production stayed the same, and the cost of the nitrogen fertilizer was US$0.06/kg, while irrigation costs US$0.02/mm, it would pay to use more fertilizer at the expense of the irrigation as the marginal cost of the fertilizer is US$0.06 relative to US$0.08 for the irrigation. Due to diminishing returns, as you apply more nitrogen in place of irrigation water, eventually the quantities will change such that at the marginal substitution times the cost is equal, indicating that the least cost way of production has been determined.

Similarly, the decision rule in deciding the best mix of products to produce is to find the combination such that the

- marginal value of each product (MVP) is identical (Decision rule 3).

This MVP is the change in physical output as one product is increased at the expense of another, multiplied by its price. Thus, for example, as an extra kilogram of wheat is produced, this might mean sorghum must be decreased by, say, 1.4 kg to release the inputs necessary to produce the extra kilogram of wheat. Now, say wheat is worth US$0.03/kg, and sorghum US$0.25/kg, then the MVP of wheat is US$0.03, and of sorghum US$0.35. Thus, their MVPs are not equal, so it would pay to produce more sorghum at the expense of wheat. As this shift occurs, the MVP of sorghum will drop, and of wheat will increase. The optimal point that maximizes profit will be where the shift is made until the MVP of each is the same. Similarly for all the other products that might be produced, when all their MVP's are equal, no change will increase profit.

With these three decision rules it is possible to determine:

- what to produce;
- how to produce it (least cost combination of inputs); and
- how much to produce per technical unit (say hectare).

Thus, any farmer who is making the basic production decisions must understand these rules and do the sums accordingly. These rules apply whether the farmer is using a linear approach to the decisions, or some other decision process. While the linear approach tends to focus on formal and significant decisions, such as whether to produce a new crop, it is also useable for simple day-to-day decisions such as whether to turn on the irrigation plant today. It is still necessary to have the relevant input–output information such as the current soil moisture, crop response to water, the weather forecast and information on its reliability, and to apply the relevant decision criteria such as whether the marginal cost of turning it on today relative to the marginal return does in fact provide a surplus, and similarly for the quantity of water applied, if any.

While optimal decision rules are clearly established, many farmers use inappropriate rules such as, for example, using average returns and costs instead of their marginal counterparts. Their conclusion will be biased.

Problem recognition

Of course, the success of the decision process does, and it is worth restating, depend on the farmer identifying the existence of a decision problem in the first place. In general, all situations are a decision problem in that *every* aspect (well most anyway) of a farm can be changed on a frequent basis. The animals can be fed more, less, the same; the field can be cultivated, or not, and so on. The trick is to first recognize when change, or the institution of something new, is a possibility, and then to work out what change, if any, is optimal.

This process of problem recognition depends heavily on the concept of benchmarks. These are standards held, normally in the farmer's mind, that provide a comparison. Thus, if the pasture, for example, is not growing by, say, 10 kg dry matter per day per hectare, then an investigation and possible action is required. On a broader scale as another example. If the wheat yield is not at least, say, 6 t/ha, then the production system needs investigating. Most farmers have these benchmarks firmly held in their mind, as does everyone else in the community, but for different systems. These benchmarks must cover every aspect of the farm against which judgements can then be made.

The big question is where the benchmarks should come from. A common approach is to survey farms in a similar environment and use the production figures from the best of these farms as the guiding benchmarks. Some would argue that this is not logical in that what is good for one farm may well not be good for another as, for example, the farmers' objectives might be quite different. Similarly, the 'best' farms might not in fact be particularly efficient anyway.

The correct set of benchmarks should, in theory, be worked out from applying production economics theory to each particular farm using the farmer's objectives. However, as this takes considerable work, the benchmarks from the 'best' farms are frequently used. This approach has the added advantage that farmers, like most people, are very keen to see what their neighbours are doing on the basis of being competitive. In the end, a simple and sensible approach is to rely on modified 'best' farm figures where adjustments are made to suit the particular farm. Thus, while the district 'best' might produce, say, 1200 kg of milk solids per hectare, the particular farm might have slightly better soil and genetically superior cows so 1300 kg might be appropriate when using the marginal return – marginal cost equality principal.

Dynamic approaches

In contrast to the linear set of decision steps many would argue that a constantly varying approach is taken by many farmers. One particular study stands out in emphasizing the dynamic nature used by some farmers. Ohlmer et al. (1998) studied a group of farmers as they made various decisions and summarized their approach with the following diagram (Fig. 4.1).

It will be noticed that four basic phases are suggested, with each having sub-phases that the farmer may use. The phases are quite broad and thus the need for sub-phases. It was suggested that the process was not linear in that the farmer might duck and dive among the various phases as the situation dictated, and as the farmer evolved his thinking.

Phases	Sub-processes			
	Searching and paying attention	Planning	Evaluating and choosing	Bearing responsibility
Problem detection	Information scanning and paying attention		Consequence evaluation. Problem?	Checking the choice
Problem definition	Information search; finding options		Consequence evaluation. Choose options to study	Checking the choice
Analysis and choice	Information search	Planning	Consequence evaluation. Choice of option	Checking the choice
Implementation	Information search. Clues to outcomes		Consequence evaluation. Choice of corrective action(s)	Bearing responsibility for final outcomes. Feed forward information

Fig. 4.1. The basic phases for dynamic approaches. Source: Ohlmer *et al.* (1998). See the acknowledgements for reprint permission details.

Ohlmer and his co-workers discovered the farmers did not necessarily work out exact figures, but used a qualitative emphasis in the analysis and choice phases. They believed that when new information was obtained, the farmers continually updated their:

- problem perceptions;
- ideas of options;
- plans and expectations.

This makes common sense.

It was also pointed out that farmers prefer a 'quick and simple' decision approach rather than a detailed, elaborate approach, and that they often conduct a small test before full implementation. This is also common sense where there is a lack of refutable evidence that a decision should be immediately and fully implemented.

Furthermore, farmers continually check clues as implementation occurs with a view to making improvements. Again, it is common sense to continually evaluate and adjust plans to suit the new conditions rather than waiting till the end of a cycle before evaluation in the cycle of:

- planning;
- execution;
- evaluation; and
- replanning

Effectively, evaluation is constant and ongoing, as is information collection. It is also related to farmers' networks and gatherings in that there is a constant need for information and thought. It is very useful for farmers to be able to discuss ideas with others in the search for a robust system, as it is to discuss proposals with family members.

Overall, it can be concluded that a dynamic approach that uses all available information as it appears with constant adjustments and revisiting phases in the total decision process is the appropriate and logical way to operate. Thus, farmers, to be good managers, need to constantly monitor, evaluate and re-plan in a completely dynamic operation. This may be uncomfortable for some in that it does require constant thinking. Some would prefer to stick to standard routines and operations. This is where the process of management is linked to management style, intelligence, and a suitable personality.

Modifications to the processes

Farmers take many short cuts in making decisions compared with the simple linear, or even the dynamic, process. They have found this necessary and desirable in seeking efficiency. For example, while it is theoretically correct to ensure all possible alternatives are considered when choosing a path to follow, this seldom occurs in reality. One reason is that a quick review of the possibilities often concludes that some alternatives do not need further analysis to rule them out. An obvious example is the removal of low-value crops requiring heavy water use from the list of possibilities when water supplies are very limited, another is growing crops that require a long growing season when the period of clement weather is very short, and so on.

In addition, using the production economic decision rules allows many possibilities to be crossed off the list before formal analysis. For example, fertilizer levels that are still in the area of increasing returns on the production function should never be contemplated, as moving into the decreasing-return range will always increase profits.

Another common approach is to find an acceptable solution in contrast to the optimal solution. The latter requires all alternatives to be assessed, whereas the former involves searching and evaluating systems until an acceptable one is found. This is quite common in many walks of life. For example, often when looking for a car to buy, a list of requirements is created. When a car that meets the list is found the search is stopped well before all alternatives are evaluated.

If the list cannot be satisfied with extensive searching, the decision maker will usually revise the list downwards until it is possible to gain satisfaction. Whether this reduced search approach is acceptable will depend on the farmer's objectives. In that it takes time to research out all possibilities there is a cost involved, and also a brain power exhaustion factor. Some would judge that eventually further analysis will not provide a net positive return.

Further to this last point, all decision processes have a cost both in time and mental difficulty. The production economics rule of going to the point of equating marginal return with marginal cost applies just as much to decision analysis as it does to production decisions. However, the value of the time will depend on its opportunity cost. If the farmer has spare time that is neither useable in an economic, nor leisure, sense, then there is no actual opportunity cost. However, as more time is put into decision making eventually the opportunity cost will become positive, as other productive jobs have to be put aside thus reducing net income. In that many farmers are very busy, a decision on how much effort to put into decision making is an important and real decision in its own right. Also it must be remembered that man does not live by bread alone, so other factors come into the choice of how much effort to put into analysing situations. In some cases pride will have an impact (being the best in the locality is a driving force for some), as will the obvious appearance of the farm relative to neighbouring farms.

Constructs

As part of dynamic decision making few farmers actually formally analyse the day-to-day decisions. Over the years they have learnt from hard experience what works, and what does not. From this experience they develop what might be called 'rules of thumb'. These are mental instructions on the best decision given specific conditions. For each circumstance and situation a different 'rule of thumb' comes into play. An example might be when to move a mob of sheep from one field or paddock to another. In early spring the rule might be that once the pasture has dropped to 1300 kg dry matter per hectare, it is time to move if the animal growth rates are to be maintained. This rule might be modified depending on the pasture levels across the farm and, in particular, in the field to which the stock might be moved. If the spring has been poor, the decision rule might be to wait till the animals have eaten down to 1100 kg dry matter per hectare.

These decision rules probably start from talking to other farmers who have more experience, or from watching a farming father. As the years progress they will be changed based on observing outcomes and seasons. Similarly, as prices change over the years they may need modifying to suit the marginal cost–marginal return rule. With higher prices it may pay to increase sheep numbers and lower the critical dry matter level. In understanding this dry matter rule it should be appreciated that the pasture intake level is dependent on the quantity of food on offer. Sheep do not simply eat the same quantity per day despite what is on offer.

When decision rules do not work in the sense that benchmark growth and production rules are not achieved, the successful farmer will start to ask himself questions on what went wrong. Technically, what is known as cognitive dissonance occurs. This refers to a feeling of knowing something is wrong. Humans adjust and move to prevent cognitive dissonance. Everyone has experienced uncomfortable situations from which decision rules are changed.

A psychology researcher (G. Kelly) formalized the idea of 'rules of thumb' with his concept of 'constructs'. He believed everyone developed constructs to aid their everyday living. These constructs evolved so that a person could deal with his outside world and feel comfortable. People might, for example, find that being polite (a construct) was the best approach to your teacher. Hard experience might have taught this lesson. Kelly had the concept of 'man the scientist' such that 'man' experiments with how to react to situations until a 'rule of thumb' (construct) is developed that he believes works well for him. Subsequently, this concept has been applied to all works of life. One example is the health field in which people develop ideas on what works for them in maintaining health. As conditions and situations change the experiments continue leading to changes in a person's set of constructs. Farmers are no different. In some situations formal analysis might occur if conditions change sufficiently, in contrast to a mental analysis for small changes. Thus, for example, when prices change markedly relative to each other a farmer might calculate budgets to compare the economics of different crops leading to a new set of rules indicating which crops to plant under different price and cost scenarios.

Intuition (or tacit knowledge), and experts

Many farmers, and people in general, make decisions without much thought, or seemingly without thought or analysis. Many an excellent farmer is said to have a good intuition in that somehow they seem to inherently know the right decision and have the skills to implement the right actions, and to make adjustments as required. This 'intuition' is sometimes called 'tacit knowledge' in the literature. This concept has been researched to a limited extent with a view to characterizing the features of good intuition. The results suggest there is nothing particularly magical about excellent intuition. People with these abilities probably have very good skills at learning from experience and observation, and somehow take on the correct constructs which are then stored for future use. Such people are probably highly intelligent and while many cannot explain the processes, they have subconsciously worked out the appropriate rules of thumb to get the job done efficiently. As time goes by, and they are exposed to the right experiences, they build up a bank of successful constructs, and they know when each should be applied.

In trying to develop good intuition it is important to practice problem-solving situations as frequently as possible, and to do so with peers and mentors enabling discussion and comparisons. Learning to ride a bicycle is probably similar. You keep trying and slowly the painful experiences leads to inbuilt systems that work; you balance, stay on and make forward progress. Gaining decision constructs is probably little different.

Part of this process may well relate to what is called 'pattern matching'. Many people work through visual images of some kind. Thus, when you see, for example, a ryegrass plant, or perhaps a white clover plant, you instantly know what it is without analysis. This is due to holding in memory a picture of the plant so that when observed the brain pattern matches and comes up with the solution. Before this pattern was stored in memory, trying to work out what a particular grass was required using first principles, or in other words, a formal analysis was required. This would consist of seeing whether the blades of grass had hairs, whether the sheaths were rounded or folded, whether the base had little ear-like protuberances, and so on. A textbook on the features of each cultivar provided the answers. With experience, this process is no longer required as the appropriate pattern is held in long-term memory. The same applies to good constructs. The problem is, of course, when conditions and situations change, further analysis is required. Thus, when the plant breeders produce a new ryegrass without the ear-like protuberance, you have to relearn the pattern which your brain instantly uses. It is remarkable just how adept it is in carrying out this pattern matching, and just how many thousands of patterns the long-term memory holds.

People who are very good at their job are commonly called 'experts'. Experts have a mass of successful constructs stored away that they have acquired over the years. As noted in an earlier chapter (Chapter 1), research has shown that experts have the following characteristics:

- excel in mainly their own area, or domain;
- successfully perceive large meaningful patterns;
- know which factors are important;
- are fast, and quickly solve problems with little error;
- have superior short- and long-term memories;
- see and represent a problem at a deeper level than novices;
- spend a great deal of time analysing problems when they are different or new;
- have strong self-monitoring skills;
- good observation skills;
- relate their judgement to the relevant objectives.

Experts must also, of course, be good at recognizing when a problem exists. They are also generally quite parsimonious over the factors they observe relative to what a novice records. Generally the learner notices and records many variables just to be sure they have taken into account everything that might matter, whereas an expert has worked out which variables are important and so simplify the problem to its essential elements.

Possibly the most important aspect of an expert is their self monitoring skill. Many people have a rich set of experiences, yet are not experts. Learning from each situation requires an ability to stand back, analyse the problem, and gather in the essential lessons. This all relates to self-monitoring skills and a realistic view of the production processes. Effectively, it is the ability to learn from experience that is important. Few people can learn from a textbook and then be an expert, unless it is something like pure maths! Thus, the right kind of aptitude and managerial style are both important, as is education.

This discussion reinforces the relationships talked about in the chapter on the origins of managerial ability, and gives the background to some of the reasons why the personal attributes are important in good management.

Both of the case farmers highlighted here believe they have developed good intuitive skills, though Margrave believes you have to be careful for conditions change so the intuitive answer can be misleading. Hank used to walk a lot and mull over the problems and decisions to be made, and he commented 'suddenly the answer was there'. In all probability, all the thinking, together with his subconscious activity, eventually produced an answer that took into account the various factors impinging, and which resulted in his 'self' being comfortable. People keep mulling if the suggested decision just does not feel right. Farmers who have the ability to get this right go on to be successful, whereas those who find 'comfort' before properly allowing for all factors end up with doubtful decisions.

Hank finds writing everything down also focuses his mind, particularly in recent times as there is less time for walking. Job lists are always a starting point to further refine decisions, which then leads to decision and action. Hank recounts the ability of his intuition over a weather event, 'we had a terrible spring and my feeling was that the autumn would be perfect, and sure enough it was and the decisions I made anticipating this situation paid off perfectly'. But, Hank points out, a bad spring is not a reason for a good autumn, it is just that in this case, 'I somehow picked up all the right signs without realizing it'.

Innovations and their adoption

New technology is constantly confronting farmers who must decide whether it is useful. While this decision is little different from any other, it is useful to list the aspects of new ideas that farmers tend to consider. In the end, of course, they need to decide whether an innovation will enhance their operation. Extension people work hard at introducing new approaches, but are often frustrated at the speed of uptake. Usually however, the slow uptake is for good reason. If farmers can see significant benefit, they will proceed with speed (as many have in domesticating deer. See Fig. 4.2).

With respect to its adoption, the aspects of a new technology that many farmers think should be considered are:

- trialability;
- complexity;
- comparability;
- observability.

Fig. 4.2. Trying out, and perhaps adopting, new ventures is always an important part of keeping ahead in the profit stakes.

If it is easy to trial the innovation without a major commitment it is easier for the farmer to come to a conclusion. Furthermore, the analysis will be much easier, as will the trials, if the innovation is not complex. Then having alternatives that are comparable provides benchmarks against which the innovation can be compared, and similarly, if the results of trials are easily observed a decision is much easier. Where the capital requirements of an innovation are not substantial, that is also a positive aspect to any change. Where all these factors are favourable, uptake of a worthwhile innovation will occur more rapidly than would otherwise be the case. This assumes the farmer believes the innovation will be of benefit to his objectives, though the 'comparability' aspect compares it with the existing, or other, alternatives.

What a farmer actually does, when confronted with a new technology, relative to what he should do in a simple rational sense can be two different things. To study this process a theory known as the 'theory of planned behaviour' (Ajzen, 1991) is sometimes used. It is hypothesized that an intention is dependent on three factors:

- behavioural beliefs;
- normative beliefs; and
- control beliefs.

The behavioural beliefs represent an individual's beliefs over the consequences of an action, its outcomes and benefits, whereas the normative beliefs are the person's beliefs about what others expect him to do. This might be the influence of a respected peer, or perhaps a spouse. In contrast, the control beliefs are the restrictions on what might be possible given the resources a farmer has, and his skill set.

Using standard questionnaires values can be given to each component leading to an assessment of the decision maker's intention which then leads to action. This theory has been applied to several agricultural situations (as well as many other industries and situations). Two examples are given by Beedell and Rehman (2000), and Sambodo and Nuthall (2009). The first looks at conservation practices, and the latter technology adoption in less-developed agriculture in which it was discovered that another important factor in intention and action is what was called a 'bargaining process'. Effectively family, and other, bargaining took place in order to come to a conclusion on action. This conclusion further reinforces the place of family in decision making which is discussed fully in Chapter 7.

Objectives and Their Impact on the Managerial Processes

Maximization

Most production economics texts would have us believe that the farmer tries to maximize his objectives through the correct choice of products, inputs and production processes. No doubt many farmers would voice this objective, but in reality it is unlikely that their actions lead to maximization. The texts also frequently assume the main objective is the maximization of farm profit where this is defined as the sum of the sustainable net cash profit and the increase, hopefully, in net asset value. Indeed, the decision rules of equating marginal return to marginal cost assumes this profit goal. In reality, as strongly noted in Chapter 3, a farmer and his family tend to have a more complex goal set, which includes factors like leisure time, pleasantness of working conditions, products that the farmer enjoys working with, protection of the environment, spending time with colleagues and friends, and so on. This is a multi-goal decision environment in which a farmer can seldom succinctly voice just what the goals are and their trade-offs.

Satisficing

In contrast to maximization, many farmers look for a satisfactory, as against the highest, level of an objective. They keep trying to improve, say, cash income, until a particular level is reached. Searching for improvements then tends to stop. Often there is a priority listing between goals so a satisfactory level of income becomes the first goal, and once it is reached, then, say, leisure time is addressed with a view to achieving at least a particular level of spare time, and so on down the list of goals. The ultimate situation is when all goals are met at the required level. This decision process involves maintaining adequate levels of higher-order goals as lower priority ones are addressed. Provided the minimum income and leisure time is achieved, then maybe the appearance of the farm is upgraded to meet the desired level. But this only happens if the higher-priority goals are met and maintained.

Between the maximization and satisficing approaches are mixed arrangements. For example, a farmer might wish to maximize income subject to a minimum total leisure time. Thus, plans are made for this minimum leisure time, and provided this is maintained the search goes on for production and systems that maximize profit.

Multiple goals

Where several factors are in the mix making up the objective, the farmer may relate the value of each to create a trade-off situation. For example, it might be considered that US$100 is equivalent to 5 h of leisure time. If the value of leisure goes up, it is worth substituting real production for leisure time where the product price stays constant. The opposite can occur if the price of the product increases, or its production efficiency so that its net return increases. These substitution rates will vary as the quantities of products produced changes. Thus, if the profits are such that leisure is sacrificed for greater profit, eventually the value of leisure will increase as the quantity taken decreases till eventually, to the farmer and his family, it is no longer worth substituting leisure for profit, and a new equilibrium is reached.

A farmer might have a value in his mind for each 'product' that might come from the farm (leisure, profit, enjoyment of particular products, enhancing the environment, etc.) so the quantity of each is shuffled round until the marginal value of each is equated; this is another production economics principle.

As most farmers do have multiple goals this mixing and matching undoubtedly occurs so that for any particular farmer there will be a mix that he aims for. Whether he tries to maximize some sort of overall satisfaction, or simply tries to reach a satisficing level of each, will depend on the farmer's personality. In this process, a knowledge of what must be given up as another goal is increased is crucial to success in coming to an appropriate solution. In some cases, of course, where there are surplus resources, it may be possible to increase one goal without decreasing another. Land, for example, might be used with more intensity to increase income without altering leisure time. This is almost a win–win situation.

Integration

Understanding that the goals held will impact on the decision processes used is important in assessing a farmer's modus operandi and, therefore, how he might improve his skills. Maximization is more difficult that satisficing, or some variant of it. It requires ensuring absolutely all possibilities are considered and demands much more time spent on data searching. Satisficing is more about searching around for a system that meets the basic requirements.

Whatever the farmer's approach, the concept of utility applies. In theory each output from a farm gives rise to satisfaction, or utility as it is called in the production economics textbooks. Usually the level of utility from production is a decreasing function, so, for example, each additional income unit, say, US$500, provides lower and lower marginal utility. The difficulty with this theory, however, is that utility does not have a comparable and definable unit. This is in contrast to, say, length for there is a standard metre against which all lineal lengths can be measured and compared. So, while in theory, everyone has some kind of internal measure of utility, it can not be compared across people.

In a multiple-goal situation each individual will have a utility value for each output, and this value will decline as more and more is produced. Conceptually, there is an equation that measures the utility from a particular farming system. Thus, for some farmer the equation might be...

$$\text{Utility} = 2Y + 4L + 1.2\,YL - 0.2\,YY - 0.1\,LL$$

where Y is dollar income and L is hours of leisure (note that this function has decreasing marginal utility due to the negative terms).

Given an equation like this, together with a knowledge of the relationships between the physical production and the output of income and leisure, it is possible to find the production system which maximizes utility, or in a satisficing situation, meets the minimum requirements at least cost.

Fig. 4.3. Leisure and relaxation are an integral part of life and must be catered for.

In reality, a farmer certainly does not sit down and work out his utility. What might happen, however, is that something like this process occurs in an intuitive subconscious operation (for example, he might well know water sports are great fun....Fig. 4.3) Somehow the farmer allows for the substitution rates and comes up with an answer. In looking at ability, just how well this happens will help define how good a manager a farmer is, and where improvements are possible.

Complications

So far no mention has been made of the vagaries of primary production. Output is uncertain, as are the prices received so it is impossible to predict exactly what the output of income and leisure might be, and similarly for the other outputs that might have utility. This means further concepts must be added to any analysis.

Achieving the goals set is often destroyed by uncertainty and risk. Accordingly, for example, a farmer might prefer a production system that can guarantee a certain minimum income. Or it could be that he will be quite happy to take the good with the bad if, on average, over several years the average profit is as high as possible. Generally, a farmer's modification to his objectives under the realistic risk and uncertainty situation is dependent on his attitude to variability. This might be a risk preference attitude, or in contrast, a risk aversion attitude, or something in between. The section 'Attitude to Risky Situations' on page 25 contains a more complete discussion on farmers' attitudes to risk and uncertainty as well as providing a question set that can be used to assess a farmer's attitude.

A risk averter is happy to give up average income in return for a relatively stable income. Such a person is probably content to take out considerable insurance, for example, in that a sure cost to maintain a reasonable income is better than a higher average income that in fact might be quite low in some years. Indeed, if a farmer's fixed debt servicing is quite high, a very poor year might bankrupt him even though the farming system used is, on average, the best possible.

In contrast, a risk preferer is quite happy to take the chance of a bad year if, on average, his income is much higher than the safe and sure approach. Each farmer will be slightly different in their attitude, and these differences probably stem from their personality and background. Also making an impact is the farmer's debt situation so if, for example, debt is low and the farmer has a good asset base a farmer is more likely to take risks.

Sorting out a farmer's risk attitude is a precursor to understanding how efficient each one is. Their attitude affects what is a correct choice, and this will be different for each farmer. Thus, one farmer might be rational in diversifying production, whereas for another diversification would be inefficient. Diversifying probably means producing a product of lower profit, but with the hope that in a poor year for one product, one of the others will be having a good year. Thus, judging managerial ability and the processes used by a farmer does depend on the farmer's objectives and his attitude to risk.

Many books have been written about optimal decisions under risk, and how to formally assess a farmer's attitude, and to calculate risk levels. It is not appropriate to repeat these procedures here, but a reader interested in this detail should consider, for example, Anderson *et al.* (1977). Furthermore, some workers believe that in many situations it is difficult to quantify the chances, or probabilities, of outcomes so a body of theory about decisions under uncertainty has been developed. (Often the word 'risk' is used when talking about situations where a chance, or probability, can be placed on each outcome. In contrast, where you know a range of possible outcomes is possible due to chance, but you cannot say whether one is more likely than others, the word 'uncertainty' is used. Tossing a coin or die, for example, is a risk situation, whereas the price of genetically engineered cotton is most likely an uncertainty situation.)

At the heart of these uncertainty situations is 'game theory' (Agrawal and Heady, 1972) in which the farmer is regarded as making decisions with nature opposing him. One of the proposed decision processes is for the farmer to consider each strategy and work out the likely income from each under representative states of nature. Then a choice is made by selecting the strategy that has the highest of the lowest payouts possible from each alternative; this is the maximum–minimum rule. Various other decision rules are possible too. One is what is known as regret theory. Here the pay-offs are listed as the regret the farmer would experience if a particular state of nature occurred for each strategy. With this data, a farmer can then select a strategy that minimizes his regret.

There is no doubt that classifying and quantifying farmers' objectives is a difficult area as the range of possibilities mentioned demonstrates. For each farmer the reality will be some mixture of the ideas covered. Anyone trying to understand how a farmer operates needs to take account of the farmer's objectives, and relate them to how the farmer might be made more managerially efficient given his objectives. A farmer, for example, very concerned about a stable income must be able to assess probabilities and how they can be used in decision making. Without these skills, whether held intuitively, or formally, he will not be a good manager. In contrast, someone who is risk-neutral need not be as skilled in risk and uncertainty analysis as these aspects are less important in their objectives.

Similarly, where a farmer has multiple goals this must be recognized and allowed for in decision making, whereas someone focused on simply maximizing average net profit needs less understanding about goal substitution and their relationships in order to be a good manager. Effectively, 'goals and objectives' is a further complicating factor in judging and training managerial ability.

Concluding Comments

All farmers are unique in terms of their objectives and goals, and in their management style and aptitude. Consequently the decision processes used will be different. In improving managerial skill it is important to understand the processes actually used, and their relationship with the farmer's objectives, if improvements are to be made.

Given that production is almost always undertaken under risk and uncertainty, a successful process is likely to be dynamic as the conditions and outcomes are always changing, and frequently in a unexpected way. Thus, the process must ensure observations are constantly being updated, and that the farmer looks well into the future when making plans. Decisions made now will be played out into varying future timespans. To make the right decision it is important to know its range of possible long-term ramifications to ensure a complete appraisal. And in making the decision, the farmer must clearly use the logical set of decision rules which have been well enunciated in production economic texts.

Farmers must, of course, be ahead of the play by recognizing when problems and opportunities arise. It must be recognized that some farmers do simply 'follow the leader' in contrast to carrying out their own analyses. If the leader is successful, and their situation is virtually the same as the case in point, the farmer is lucky especially if his implementation skills are up to the task. This is the exception, so farmers must be aware of the outcomes and efficiencies that are possible if they are to be good managers. 'What is possible' gives rise to the benchmarks.

Implementation of the decision processes must recognize the farmer's specific objective set, and appropriate modifications to standard procedures put into effect. Especially if the farmer is other than risk-neutral, due allowance must be included in analyses that take into account variability. And where there are multiple goals, calculations, whether formal or mental, must allow for each objective thus complicating the process compared with simple profit maximization.

In working to improve the processes that make up managerial ability, it is important for farmers to discuss the issues with mentors of various kinds. These might include family members, but also farmer friends. It is through constructive criticism from self and colleagues that processes are improved. The objective is to become an expert with the associated attributes. Practice, and discussion, helps create the skills necessary. Success creates rewards which reinforces the developing process, and vice versa. Eventually, the result is a permanent change in behaviour.

In that goals tend to be dynamic, with maturity and experience people change what they want from their business life, so an optimal decision process is also dynamic. Sometimes feelings change from day to day, so care must be taken in sorting out a robust process that serves well over time. Furthermore, when it is clear that some goals are impossible to achieve, people rationally lower their sights somewhat to create practical aims and outcomes.

In a nutshell, success depends on:

- recognizing problems;
- making the correct observations;
- successfully applying the correct decision rules;
- successfully implementing the decisions made; and
- successfully monitoring and adapting the decisions in response to the ever-changing situation.

5   Skills Required

Introduction

Any farmer must have as a minimum certain skills, but for excellence the farmer must be good at the complete list of competencies, or skills, necessary to operate a farm. There are various opinions on the skills that make up the list, and no doubt this varies to a certain extent with the type of farming, but there is certainly a core set common to all situations. You can probably quickly come up with a list of what is necessary after some thought. A successful farmer must have, first,

- good technical skills as without them the jobs will not be successfully completed.

Similarly, a good knowledge of the technology is essential. For example, understanding how pasture might respond to fertilizer application is essential for correct decisions. There are a myriad of such technical 'facts, figures and relationships' that are a prerequisite to good decision making, as is the ability to apply the decisions.

Other examples of the skills required include:

- observation ability;
- visualization, or prediction, of likely outcomes resulting from any decision;
- negotiation skills;
- people relationship and management skills;
- risk management skills;
- an ability to simplify situations down to their essential elements; and so on.

This chapter contains an outline of just what these generic skills are likely to be. This is achieved by reviewing survey information seeking farmers' views of what skills are important, and then assessing their views, and adding to them as required.

Whilst having an appropriate aptitude, personality and experience are all essential, in a specific sense all these attributes must be brought to bear on each of the skills required. In the end, it is the level of the skills that determines the outcomes with the basic building blocks being the precursors to successful training in each of the skill areas. Training, of course, is often an 'on the job' process in contrast to formal course work.

This chapter proceeds with the results of the farmer survey, and then a discussion on the skill sets. Some of the data presented and discussion is based on Nuthall (2006). The chapter finishes with an introduction to two professional consultants and their views on the required skills for farmers. The views of these two consultants on training methods, and on ways farmers can improve their skills, particularly with respect to the place of consultants, are presented in Chapter 8.

The Questionnaire, Sample and the Respondents

The questionnaire was developed from the literature on competencies for a range of countries, and discussing the possibilities with farmers and agricultural consultants. The questionnaire used is presented in Appendix 5A. The potential competencies were broken down into three groups:

- managerial attributes;
- entrepreneurial skills; and
- personal attributes.

This division ensured the set to be scored was kept to manageable proportions. The distinctions were somewhat arbitrary.

The random sample of 2000 farmers (which was approximately 5% of the total population) was divided into:

- 16 statistical regions;
- six farm-type groupings (intensive and extensive sheep, cattle, deer, dairy, cropping and horticulture); and
- 12 area (hectare) groups.

The number selected from each group was based on the proportion of the total population in the group; 823 usable responses were obtained giving a response rate of approximately 41%.

In order to put the data that follows into perspective relative to a range of countries, details of the farms and farmers are presented. Tables 5.1–5.3 give the farm type, labour used (including the manager) and land area distribution of the respondents. In cases of mixed enterprises the farm was classified by the major enterprise.

It is clear one- to two-person units dominate and that while dairying is increasingly important, more extensive grazing properties involving sheep, cattle and deer (45.5%) are the most numerate. However, there were large numbers of smaller-sized properties with dairying and horticulture probably being

Table 5.1. Distribution of farm types in the sample (% of total).

Intensive sheep	17.5	Dairy	33.4
Extensive sheep	12.0	Cropping/horticulture	16.6
Cattle	12.7	Other	4.5
Deer	3.3		

Table 5.2. Distribution of labour used
(including the manager) (% of total in each category).

Number of units	Percentage
≤1.0	24.0
1.1–2.0	47.1
2.1–3.0	13.5
3.1–4.0	7.2
4.1–5.0	4.0
5.1–6.0	1.0
>6.0	3.2

Table 5.3. Distribution of area (hectares) used by the
respondents (% of sample).

Area range (has)	Percentage
≤50.0	20.2
50.1–100.0	16.5
100.1–150.0	11.6
150.1–200.0	8.6
200.1–250.0	6.4
250.1–300.0	6.8
300.1–350.0	3.2
350.1–400.0	4.0
400.1–450.0	2.1
450.1–500.0	2.5
500.1–550.0	2.1
550.1–600.0	2.3
600.1–650.0	1.6
650.1–700.0	0.7
>700.0	11.4

the dominant uses. For the under 100 ha class, dairying makes up 37.5% of the farms, cropping and horticulture constitute 32.7%, cattle 11% and 'other' 8% of the units, leaving deer and intensive sheep to make up the remainder (10.8%).

Table 5.4 presents farmers' education. It shows the farmers are reasonably well educated so they have probably thought carefully about what competencies are important.

Table 5.4. Distribution of formal education levels.

Percentage reaching the following levels	
Primary	2.3
Secondary – up to 3 years	35.7
Secondary – greater than 3 years	28.6
Tertiary – up to 2 years	13.6
Tertiary – greater than 2 years	19.5

Managerial Attributes

Respondents were asked to rate the importance of a range of attributes on a 1 (not at all important) to 7 (very important) scale. Table 5.5 gives the results for both the farmers and a sample of professional agricultural consultants (approximately 320).

An analysis of variance showed the differences between the means was highly significant with $p = 0.0$ ($F = 199.05$ and 236.45 for consultants).

While the list is ordered according to the farmers' ranking, the ranking according to the consultants is given in the brackets. The order changes slightly, but generally the two groups agree on what is important.

The three most important attributes were:

- observation;
- introspection (key factors and priorities); and
- communication.

These embody the four highest-ranked attributes for both groups of observers.

Eight of the 15 items are scored 5.5 or better indicating many attributes are considered important. This might be expected as only the most likely were included in the list offered. To help analyse the responses a factor analysis was carried out. This looks at the correlations between all the items to isolate the groups that tend to go 'hand in hand'. Studying the components of each group might well suggest some basic attributes that underlie those listed. Table 5.6 contains the results of the factor analysis.

These two underlying factors explain 45% of the total variance. Note that factor loading values less than 0.3 have not been presented as they contribute in only a minor way to the factors. The data is interpreted through noting that, for example, Factor One is made up of:

- 72% (or 0.72) of item 9 (making requirements clearly understood);
- 68% of items 1 and 15 (identifying key factors and assessing job priorities);
- 62% and 61% respectively of items 11 and 2 (knowing how to choose between alternatives and quickly sorting out new situations); and so on for the rest of the items.

In Factor Two the most important item is:

- at 80% 'understanding the local political scene'.

Table 5.5. Importance of managerial attributes. Mean scores on a 1–7 scale (ranging from not important to very important).

	Farmers	Consultants	(Order)
1. Being up to date with the current condition of the property in its totality (bank balances, animal condition, crop growth, soil moisture, feed levels, machinery repair)	6.23	6.07	(3)
2. Ability to identify the key factors in a problem, and discard the irrelevant	6.16	6.29	(1)
3. Making requirements clearly understood (effective communication)	6.13	6.28	(2)
4. Assessing job priorities	5.93	6.00	(4)
5. Quickly analysing and sorting out situations that have never been faced before	5.68	5.26	(12)
6. Having a clear understanding of the family's objectives, values and goals, thus making assessing the value of alternative actions easy	5.67	5.79	(5)
7. Picturing (understanding) the consequences of a decision over the many (or few) months/years it might impact over (e.g. planting an area in forestry, subdividing a paddock)	5.63	5.71	(7)
8. Being able to efficiently organize and carry out quite complex operations (e.g. get a new packing shed operational on time)	5.61	5.52	(8)
9. Developing appropriate and detailed plans for both short- and longer-term horizons	5.47	5.71	(6)
10. Understanding the basis on which to choose between alternatives (e.g. knowing how to cost unpriced labour, knowing how to do gross margins, understanding diversification principles)	5.31	5.32	(11)
11. Skill at keeping, interpreting and using recorded data about the property and associated factors (e.g. market trends)	5.17	5.42	(10)
12. The ability to predict product prices into the foreseeable future, or at least understanding the factors that determine the prices, and understand market requirements	5.16	4.96	(13)
13. Developing and maintaining a support network of colleagues and professionals	4.89	5.44	(9)
14. Being able to predict local weather better than the official forecaster	4.23	3.07	(15)
15. Understanding the local political scene as it might impact on rules affecting what can be done	3.88	3.40	(14)

Table 5.6. Factor analysis loadings (item contributions) for
the managerial attribute variables (Refer to Appendix 5A
questionnaire list for the attribute represented by each number).

	Factor number	
Attribute number	ONE	TWO
1	0.68	
2	0.61	
3	0.42	0.31
4		0.66
5		0.80
6	0.30	0.53
7	0.47	0.38
8	0.56	0.38
9	0.72	
10	0.54	0.45
11	0.62	
12	0.58	
13	0.48	0.51
14	0.33	0.62
15	0.68	

However, this item has a low ranking (15th), while it is a major contributor to the factor this does not necessarily mean the factor as a whole is important. Indeed, if the scores of each factor component are averaged, Factor One is rated 5.62 and Factor Two is 4.99, indicating the importance of Factor One.

While the farmers believe there are two basic sets of attributes that go together, the consultants believed there were four groupings (with average scores of 5.66, 5.46, 5.84 and 3.97). Clearly the consultants believed managerial skill was made up of more factor components than the farmers. However, the items left out of Factor One by the farmers were also left out by the consultants. Given that the score rankings were much the same for both groups, the conclusion must be that the most important attributes have been clearly stated.

The above analysis has grouped all farmers together no matter their farm type or personal attributes. Consequently, the analysis was repeated for a wide range of different groupings including:

- farm type;
- age;
- gender;
- managerial style;
- education;
- self-assessed intelligence;
- managerial ability;
- objectives; and
- whether a farm computer is used.

Generally it was concluded that the results were largely the same no matter which groupings were compared. Farmers, therefore, had the same views no matter which farm type, location or characteristic set they belonged to.

Entrepreneurial Skills

Respondents were asked to rank 12 entrepreneurial skills on a 7 (very important) to 1 (not at all important) scale. They were also given the opportunity to add further skills if they thought those offered did not cover the full list of possibilities. While some respondents did write in their thoughts on additional skills, most were a rewording of the 12 listed skills. Table 5.7 lists the scores given on a 1 (not important) to 7 (very important) scale ranked in order of descending importance. The scores given by the consultants are also presented.

The F test showed the difference in the means was highly significant ($p = 0.0$, $F = 49.48$). The five most important skills involve:

- meeting deadlines;
- successfully obtaining decision information;
- price negotiation;
- successfully handling risk; and
- an anticipatory intuition.

These skills are scored very similarly except for the top priority of meeting deadlines – this stands out.

The consultants have a similar priority list so there is general agreement, though the 'ability to learn new skills' is ranked lower by the farmers, as is a belief in being able to control what happens. Clearly consultants believe farmers still have some learning to do!

As for the managerial attributes, the farmers believe a wide range of skills are important. Nine of the 12 listed skills are ranked 5.55 or greater. The lowest ranked, by both the farmers and consultants, was the ability to forecast longer-term opportunities – perhaps they do not believe there will be new and promising opportunities and that improving on existing systems and products is more relevant.

To examine whether farmers believe there are inherent groupings among the skills a correlation (factor) analysis was conducted. Table 5.8 presents the results (loadings less than 0.3 are not presented due to their insignificance).

Compared to the consultants, the farmers saw entrepreneurial skills more simply in that they grouped them into two factors (that explain 56% of the variance) instead of three (explaining 59% of the variance). Interestingly, the important skills in Factor One are:

- information seeking (0.75);
- an ability to learn new skills (0.75);
- dealing with risk (0.72);
- an ability to look ahead (0.66);
- a full comparison of alternatives (0.64);
- early warning sign intuition (0.63); and
- a belief in being able to control many factors (0.61).

Table 5.7. Importance of entrepreneurial skills. Mean scores on a 1 (not important) to 7 (very important) scale.

	Farmers	Consultants	(Order)
1. Understanding deadlines and being able to 'act in time' (e.g. spray before insect damage, fertilizer applied in good time)	6.16	6.38	(1)
2. An ability and determination to look/ask/seek out information thought to be necessary for making decisions	5.78	5.99	(2)
3. The skill to negotiate the best possible deal (price, arrangement)	5.78	5.34	(9)
4. Understanding sources of risk and what can be done to reduce its impact	5.75	5.70	(4)
5. An intuition that gives early warning signs when something is not right, or, in contrast, when something positive needs exploiting	5.75	5.65	(6)
6. Ability in learning new skills	5.58	5.71	(3)
7. An ability to look ahead and anticipate likely problems, needs and opportunities	5.70	5.61	(7)
8. When faced with opportunities, ensuring that ALL alternatives are sought out, considered and evaluated	5.65	5.35	(8)
9. A belief in being able to control a lot of what happens around the property in contrast to a belief that not much is really controllable due to the weather, markets or government action	5.55	5.69	(5)
10. Skills in finding the very best market (price, quantity, etc.) for all output	5.34	5.03	(11)
11. Being able to seek out, identify and clarify new opportunities (production, products, marketing, etc.)	5.21	5.25	(10)
12. The skill and intuition to forecast well into the future likely opportunities in products and production systems.	4.90	4.68	(12)

Table 5.8. Factor analysis loadings (item contributions) for the entrepreneurial skill variables (refer to Appendix 5A questionnaire list for the skills represented by each number).

	Factor number	
Skill	ONE	TWO
1	0.35	0.62
2	0.75	
3	0.75	
4	0.63	0.32
5		0.88
6	0.50	0.39
7		0.76
8	0.61	
9	0.64	
10	0.39	0.68
11	0.66	0.35
12	0.72	0.32

All these skills are seen as a connected 'kit bag' involving common sense data collection and analysis, and a perceptive observation system that is tuned to opportunities and problems.

The important components of the second factor are skills in:

- marketing (0.88);
- negotiating (0.76);
- forecasting intuition (0.68); and
- an ability to discover new opportunities (0.62).

These are clearly connected. Which factor is more important? Factor One has an average score of 5.60, and for Factor Two it is 5.57, so both must be regarded as significant.

With respect to sub-groupings based on variables such as farm type, while there were some statistically significant differences between some of the mean scores for the different groups, the rankings of the various skills only changed marginally.

Personal Attributes

The questionnaire included 18 personal attributes that the respondents were asked to score on a 1 (not important) to 7 (very important) scale. Table 5.9 lists the attributes in score order. An analysis of variance indicated the differences in the mean scores were highly significant ($p = 0.0$, $F = 207.20$).

It will be noted from the table that seven of the attributes have a score greater than 6.0, and another three are greater than 5.7. Overall, the respondents have

Table 5.9. Importance of personal attributes. Mean scores on a 1 (not) to 7 (very important) scale.

	Farmers	Consultants	(Order)
1. Early observation of important indicators around the property (e.g. lambs are scouring, wheat is infected, cows losing weight, pasture growth has increased, etc.)	6.65	6.72	(1)
2. Ability to learn from experience, mistakes and failures	6.35	6.28	(2)
3. Developing a 'good moral character' involving openness, integrity, reliability, trustworthiness, etc.	6.35	6.10	(3)
4. Maintaining good relationships with outside people – bankers, accountants, suppliers, etc.	6.19	5.87	(6)
5. Keeping a cool head and putting aside any tendency to panic when faced with stressful situations	6.19	5.79	(7)
6. Having the confidence to draw conclusions and act quickly and decisively	6.18	5.95	(4)
7. Obtaining employees' and/or contractors' cooperation and understanding leading to harmonious and productive relationships	6.08	5.91	(5)
8. Understanding the interrelationships between all the components of the property (e.g. rainfall–soil moisture–plant growth–animal grazing, i.e. what affects what?)	5.99	5.77	(8)
9. Successfully resolving conflicts on, and off, the property (e.g. dispute between employees)	5.78	5.57	(10)
10. Successfully judging personality and selecting suitable employees	5.74	5.53	(11)
11. An excellent knowledge of facts, figures, procedures and methods, with respect to soils, plants, animals, machines, buildings, etc.	5.58	4.99	(12)
12. Accepting the good and the bad and not letting it affect management and decision making	5.53	4.93	(13)
13. High motivation in constantly seeking better ways and implementing them; in contrast to being happy with current systems	5.28	5.75	(9)
14. The determination to keep working all hours, until the high-priority jobs are completed	5.24	4.48	(15)
15. Being prepared to give it a go and take risks in changing production systems and/or starting new ventures	5.14	4.84	(14)
16. Developing a strong personality so that others 'sit up, notice, respect, and act' on what is said	4.96	4.27	(17)
17. Tertiary education in areas related to primary production (agriculture, horticulture, biology, marketing, etc.)	4.61	4.33	(16)
18. Having above-average intelligence and school grades	4.46	4.19	(18)

scored the personal attributes rather higher than for the managerial attributes and entrepreneurial skills. Clearly, early observation of important indicators is regarded as the top attribute or skill, and the ability to learn from experience and developing a 'good moral character' are not far behind.

Maintaining good relationships with business associates outside the farm and 'keeping cool' under all circumstances are also scored among the top attributes. Being able to act decisively and good relationships with employees or contractors are also highly rated. At the other extreme, high intelligence and good school grades, and developing a strong personality, are not regarded as being particularly important. Similarly for a tertiary education in areas related to primary production. Yet, it must be noted that approximately one-third of the respondents had experienced tertiary education, though the areas of study are not known.

With respect to the attribute rankings, it is interesting to note that at least in the area of personal attributes the consultants are virtually in agreement with the farmers, both in the ranking and some of the score levels (highest is greater, but lowest is lower).

To assess the groupings of the attributes a further correlation (factor) analysis was conducted. The loadings for values greater than 0.3 are given in Table 5.10. The three-factor solution (with eigenvalues ≥ 1) explains 53% of the variance.

Table 5.10. Factor analysis loadings (item contributions) for the personal attributes (refer to Appendix 5A questionnaire list for the attributes represented by each number).

Attribute number	Factor One	Factor Two	Factor Three
1	0.50	0.50	
2	0.54	0.50	
3	0.58	0.51	
4		0.46	
5		0.70	0.32
6		0.71	0.34
7	0.34	0.54	
8	0.58	0.48	
9		0.38	0.33
10	0.68		
11			0.60
12	0.49	0.45	
13	0.70		
14			0.69
15			0.76
16	0.73		0.32
17	0.75		
18	0.70		

Factor One is all about personality and relationships:

- developing and maintaining good working relationships both within and outside the property, the ability to learn from experience;
- early observation;
- a personality that does not panic; and
- acting quickly when required.

Perhaps 'early observation' is out of place as a cohort, but possibly it is related to a personality that is careful and gives attention to detail.

Factor Two has as its important components what might be called:

- an adventurous spirit ('give it a go', take risks, keenness to try new ways, and so on) as well as;
- early observation;
- acting decisively; and
- not panicking.

The synergies here are clear.

The important components of Factor Three are above-average intelligence, tertiary education and a strong personality. All these factors are not regarded as being relatively important, but they clearly relate to each other, or at least the first two do. The average score of the components of each factor is 6.09, 5.81 and 5.06. The first two factors dominate in importance.

As before, the rankings remain relatively stable when the full ranges of sub-groupings are compared.

Bringing Together the Farmers' Views on Competencies

The data as presented does not provide very distinctive and clear-cut conclusions on a short list of the most important competencies. A simple list of, say, six to ten competencies with scores well above the rest did not emerge. The respondents believe there is a wide range of skills that are part and parcel of managerial success, and any line used to divide an important and less important group will be at an arbitrary position. It was therefore important to look for correlations between the higher-ranked competencies to see if the members of the groups have similarities than can be used as core factors in training programmes – thus the various factor analyses that were presented. To further enhance the factor groups, the competencies from all categories with a score greater than 5.69 were grouped and reanalysed. Table 5.11 gives the results. Using 5.69 as the divider provided 18 items with a factor loading greater than 0.5.

The factors explained 54% of the variance with a very high level of significance ($p = 0.0$).

It is clear Factor Three is about good skills in selecting and managing people. Factors One and Two are more complicated and involve several competencies. Factor One is about planning and associated issues such as

Table 5.11. Factor analysis loadings of competencies from all groups with a score greater than 5.69[a].

Competency (paraphrase)	Factor One	Factor Two	Factor Three
Observing current state of farm		0.57	
Planning for short and long terms	0.52		
Obtaining planning information	0.59		
Intuitively noting early signs	0.64		
Acting on time	0.65		
Negotiation skills	0.65		
Looking ahead and anticipating	0.71		
Good risk management	0.73		
Early observation of important factors		0.69	
Keeping a cool head		0.66	
Confidence to conclude and act		0.62	
Learning from experience		0.63	
Developing a good character			0.57
Understanding interrelationships		0.56	
Getting cooperation of employees/contractors			0.59
Successful judge of personality			0.77
Resolving conflicts			0.80
Good relationships off the farm			0.64

[a]Only loadings of 0.5 or greater are displayed.

information-gathering and risk management. It is also about effective implementation of the plans:

- looking ahead and anticipation;
- intuitively picking up important signs;
- successful negotiations; and
- acting on time.

Effectively Factor One is about planning, implementing and control. Factor Two is similar and really reinforces the implementation component of Factor One through:

- early observation skills;
- keeping a cool head;
- confidence to decide and act quickly;
- learning from experience; and
- understanding all the interrelationships between the components of any system.

These factor analyses make the farmers' views of the components of good management very clear.

Furthermore, these components are relatively stable across different sectors of the primary producing industry as it was shown the rankings change very little with age, education, farm type, managerial style, gender, profit objective variations and computer ownership.

Summary of Skills Required

Introduction

Other than the skills rated by the farmers, some thoughts suggest further skills likely to be important in good management. As these slot into the rated list, the farmers' list will be summarized with these additional skills introduced. These are not presented in any kind of priority order for they are all important core skills, though in each farm situation some will be more significant than others.

Risk management

First, RISK MANAGEMENT is recognized by the farmers as being important. Risk and uncertainty is certainly a dominating factor within agriculture in most countries due to the weather and markets in particular. Every farmer must be aware of the nature of the risks, and what can be done about alleviating the problems. Just how much alleviation will clearly depend on the farmer's objectives and attitude to risk as previously discussed. Selecting the right alleviation systems to suit the farm and objectives is, therefore, an important skill. This means being fully aware of all the strategies such as diversification (which can take many forms. See Fig. 5.1), selection of low variability production processes (e.g. irrigation) and products, use of contracts and forward selling, insurance, flexible systems and products that can be changed to suit the situation, and so on.

Fig. 5.1. Diversification can be an important way to reduce risk. Farmers must use their imagination to come up with solutions.

Observation

Second, and perhaps before risk management, comes OBSERVATION as noted by the farmers. They specifically talked about noting the early signs of relevant situations so that action can take place in good time, and acquiring accurate planning information through observation both on and off the farm. This observation must also encompass problem recognition.

Not talked about by the farmers was the skill in knowing what to observe. Many farmers keep myriads of bits of information which are never used. Thus farmers must be able to sort out the relevance of observations and discard the useless. This involves critical thinking. That is, the skill to look at information and decide whether it is accurate and relevant. How often have you been told in an advertisement, or popular article, that, say, a particular soil additive is absolutely essential for good growth. If you believed all advertisements in their entirety, you would be buying a significant heap of inputs that would put you on the wrong place of the production function. Critical thinking should be a farmer's constant companion.

Good observation requires:

- good listening;
- reading; and
- watching skills.

Sights and sounds constantly bombarding a farmer must be noted, sorted, discarded or stored as the case maybe. How good are you at being shown a table full of objects, and then remembering what you were shown? There are many books discussing how to improve your observation skills, so every farmer should hone these attributes. But they must be good at not only visual observation, but also at picking up the relevant information when listening to people in one to one situations, or group situations, and also at picking up the relevant information when reading.

Listening skills do vary considerably, and can be learnt through a good text. While listening sounds like a passive affair, a good listener is actually practising what is called 'active listening'. This primarily involves feeding back to the speaker little summaries of what they have said to ensure you have picked up the central message. Encouraging, but brief, comments are also important to show the speaker that you are noting what is said, and are supporting his or her efforts.

Similarly, reading skills vary between people. Research has shown a particular set of approaches is beneficial. This involves:

- skimming to start with so you can decide whether the material is relevant; and
- then starting back at the beginning picking up the main messages.

A good memory helps with these skills for you want to be able to store and retrieve the important and relevant information, or, at least, the essence of it and where the details can be found when needed. Again, there have been many books produced on how to develop memory skills.

No matter how good the memory, a farmer will usually need some kind of recording system, even if just for tax purposes. Thus, the skills to:

- know what to record; and
- how to do this

are important. Essentially, records should be kept if they can be used for improving decisions provided the cost in time of obtaining and recording the records is not greater than the increased return from their use. That is, information has a cost and return so the farmer should keep sufficient records such that the marginal cost equates with marginal return. Of course, the skill is in knowing this point! In the record-keeping process, the use of a farm computer may be relevant provided adequate integrated software is available. But remember some records will be kept for other than monetary reasons. With a strong interest in the environment, perhaps a farmer keeps records of wild fowl activity purely for general enjoyment.

Finally, under the observation skill, a farmer must always be on the lookout for new knowledge that might have some application on his farm. This will involve:

- extensive reading;
- talking and listening;
- field day visits;
- computer searching and watching.

Indeed, one assessment of the features of successful managers showed the one common thing between them was the time spent reading and searching out new knowledge, be it technical or financial.

Negotiating

Third, the farmers rated NEGOTIATING skills highly. Most farmers will indeed be involved in many negotiations, from fixing prices through to setting contracts for new employees. Special skills and training are important to good negotiating. Some people spend their life as professional negotiators. Courses and texts on negotiation abound.

One of the central themes is that there is almost always a set of conditions that both, or more, parties will be content with. The skill is in using imagination to come up with a suitable package. A simple example is the farmer who wants a certain price for his wheat, but the buyer is offering somewhat less. Clearly both may need to alter their stance, but by how much depends on the competitors positions. Nevertheless, perhaps there are other factors that will help. The farmer might have some storage facilities which can be used if the buyer is struggling to keep all the grain he needs. So, a compromise might be a small drop in price together with the farmer agreeing to hold the wheat on farm for 2 months. A down payment might also be arranged.

In negotiations it is important to consider all factors that might impinge on the situation and work around variations on these until a meeting of the minds occurs. Of course, in some cases it will pay to walk away from a negotiation where the seller recognizes that someone else can offer a better deal.

Anticipation

Fourth, ANTICIPATION is a critical core skill, as recognized by the farmers. Anticipation of what might occur is critical in ensuring all situations can be catered for. If opportunities are lost, or problems not corrected, the outcomes will be less than what is possible (in succession planning (Fig. 5.2) delay may destroy opportunities). Anticipation is particularly important when planning because a farmer must imagine what the outcomes might be from following certain decision paths. The whole of decision making revolves around antici-pating what might happen as a result of taking alternative decisions. Without successful anticipation decisions become a haphazard affair.

Successful anticipation depends totally on visualization. The farmers, most of whom tend to be 'concrete' thinkers, must be able to mentally visualize what might happen this month, this year, next year and perhaps through to several years, as a result of taking possible decisions. The results of the visualization lead to calculating comparative budgets, even if mentally, giving the summary figures enabling decisions. An important component of visualization is imagination.

Fig. 5.2. Planning ahead is critical. Anticipation and succession planning must start very early.

It is well known that some people have more active imaginations than others, but there are training packages that can help bolster imagination.

It is also interesting to note that most successful sportspeople are encouraged to use visualization, and there is clear proof that those that are good at it succeed. A goal shooter visualizes the ball arching away from his boot and entering the goal just out of the reach of any defenders. The high jumper has a mental picture of where his limbs will be as he clears the bar. The mental image is a bit like the TV replay, but in this case it is before the event.

Planning

The farmers also rate PLANNING FOR THE SHORT AND LONG TERMS (fifth) as being important. This must involve making economic forecasts and comparisons that involve the visualizations of inputs and outcomes. However, after the decision on which alternative path to take, the plan then becomes the blueprint for action. What has to be ordered? And when? As without the inputs on hand at the right time, and the means of implementing the tasks, success will be less than optimal.

Learning from experience

Sixth, LEARNING FROM EXPERIENCE was a skill the farmers thought very important. There is unlikely to be any industry where the ability to improve management skills from the lessons of experience is not important. Everyone has to start somewhere. Textbooks and courses are not a substitute and cannot provide an appreciation of all the factors involved in the decision making, implementation and control cycle.

Despite the importance of experience, little research has been conducted on the factors important in gaining experience. The intrinsic skill of self-observation is probably important. To learn you need to analyse what was carried out, what went wrong, or succeeded as the case might be, and how improvements might be made for future situations that mirror the experience. However, it must be recognized that seldom do exactly the same situations repeat themselves in primary production, so learning from experience will involve the ability to intelligently note the differences of the current situation, and modify the experience of the past to suit. Recognized experts are people that have learnt from their experience. There is also some luck in experience in that one farmer may have had all the right experiences in the past in contrast to another that may never have experience, say, a season that is twice as wet as normal.

People skills

PEOPLE SKILLS is undoubtedly an important attribute, and this was clearly recognized by the farmers. As the seventh core skill the farmers believed successful judging of personality, resolving conflicts, and good relationships off the farm were all important aspects of 'people skills'. People skills are also no doubt related to negotiation skills. In that a farm operates in an environment of people, being able to communicate successfully is all-important, as is a tolerance of different views and ways, provided the job gets done. Being able to maintain friendly relationships with neighbours, consultants, bankers, and so on is crucial to success. Again, there are many books and training courses on people skills that can help develop these skills.

Implementation

Eighth, IMPLEMENTATION OF PLANS is a critical skill as recognized by the farmers.

Optimal plans are no use if they cannot be successfully implemented. These two links in the chain must be as equally strong. As you would expect, the farmers commented that 'acting in time' was critical, as was 'not panicking'. Research has shown that farmers experience considerable stress relative to other occupations due to the lack of control they have over many of the conditions. Given a series of bad outcomes it is important to maintain logical decision making despite the pressures. In this respect, a farmer's personality is an especially important factor. 'Acting in time' requires little comment – in a biological world processes cannot be turned on and off at will. The same applies to markets. It is clear that it is the early bird that catches the worm on both accounts.

To this might be added the importance of 'doing the right thing here and now'. The whole primary production process is dynamic and fluid requiring constantly changing action to suit the circumstances. Equally, decisions should not be made ahead of time and rigidly maintained for they may no longer be appropriate as, for example, the soil moisture is different from the expected. A farmer must have a range of strategies in his 'kit bag', each one of which is used according to the particular condition set. Waiting to the last minute to see the conditions is equally as important as not delaying action. It is a fine balance that makes successful primary production such a fascinating study and challenge.

Technology

The farmers also recognized that UNDERSTANDING THE TECHNOLOGY was critical in that they rated highly 'understanding interrelationships'. This ninth core skill is patently obvious for without it decision making becomes nothing but a lottery. To this must be added TECHNICAL SKILLS for without these implementation becomes a rather hit and miss affair. In that primary production involves many disciplines in all their complexity, the art of simplification becomes very important. It is essential to be able to isolate the important and essential components of the technology and its interrelationships to reduce the decision problem to manageable proportions. It will be recalled a feature of experts was their ability to simplify, in a meaningful way, the complexity of a problem.

Solutions

What the farmers did not mention at all was the skill of creating realistic solutions to problems. This might be called the SOLUTION GENERATION skill (tenth). It is all very well to recognize problems and opportunities, but to create the optimal solution, particularly where it is new and unique, is rather more difficult. This skill requires an intimate understanding of the technology, markets and production conditions as well as creativity and imagination. Many farmers lack the latter two attributes, largely because they have not been encouraged to be creative. There are training programmes which promote thinking of new solutions and processes.

Analysis

Finally, the eleventh core skill is ANALYTICAL KNOWLEDGE. The farmer can be very creative, but suboptimal outcomes will eventuate if the farmer does not understand how to analyse the alternatives both in a tactical and a strategic sense. An outline of production economics principles was given earlier. These must be used if the right choices are to be made. A knowledge of what constitutes a fixed cost relative to variable costs is essential in deciding action, as is the concept of 'opportunity cost' in which it is necessary to use the value of a resource in its next best use when looking at the return from an action which takes away resources from other uses. This assumes that it is possible to measure in some way the 'farm's' objectives because this measurement must be the yardstick used. This is difficult, so the practical approach of using monetary costs and returns is often used with subsequent subjective adjustments for other objectives. Thus, the alternatives might be ranked on their financial returns, and then possibly re-ranked once their contribution to leisure time, enjoyment, complexity and so on is subjectively taken into account.

With the development of management software, computer packages are becoming available for helping ensure that the correct analytical approach is used. Similarly, packages that will determine practical optimal courses of action are slowly appearing. There are also many packages that help keep records which can turn data into information, the distinction being that information can be used directly for decision making. For example, a financial package that records all transactions, if properly designed, can produce enterprise profits which might then be projected into the future after allowing the farmer to update the past information to reflect the outlook.

Case farmers' views

The case farmers struggled a little when asked what they believe were the most important skills. They probably just do not think in these terms, preferring to just 'get on with the job' for making decisions does not require them to make such lists. Margrave certainly knew which decisions he found the hardest, even though in general he noted 'I have never found making decisions particularly hard...they just got made'. For the rare difficult decision he was not comfortable with, and spent some considerable time thinking about, once made he simply moved on to the next problem. In general, he believed staffing decisions were the most difficult, perhaps because humans cannot be quantified as easily as other factors. 'Appointments, sackings and correcting staff' were all difficult. One suspects that most managers feel the same as on a farm you work intimately with the staff.

Generally, Margrave believed important skills and attributes were:

- knowing when to act;
- having a strong desire to be a good manager, to strive to improve;
- being able to enjoy working with people;
- good communication skills;
- having an enquiring mind;
- an ability to ask someone else for another opinion, or for information and ideas;
- an ability to constantly watch for new information ('awake enough to tap into the information available');
- good anticipation skills, like constantly doing feed budgets to anticipate feed deficits and surpluses; and
- credibility with labour. The manager must be able to perform all tasks successfully so the staff respect and admire the manager.

Having provided this list Margrave also commented: 'good profits are almost a chance factor...if no floods occurred, or a drought didn't arrive, then profits were good as the farm was normally producing at high levels. Sudden upswings and downturns in the market also had an impact as in such cases forecasting was impossible. So often the political decisions on the other side of the world impacted.'

When talking about staffing issues, Margrave wanted to make the point that it was very important to work with staff and give them responsibility as well as opportunities to take off time to attend short courses. Each staff member had his own area of responsibility and was generally left to get on with that area. The non-threatening approach Margrave took when helping and watching over outcomes seemed to be appreciated by the staff. Staff turnover was minimal, reflecting the success of Margrave's approach and management. Margrave believed 'most people want to do the right thing', so providing training and praise for good outcomes gave rise to a contented and successful set of workers who more than repaid the time and money spent on them.

Overall, when asked which skills were important to his success his reply was 'I'm successful because I enjoy farming and the manager's job. I was not much use at school, but the moment I left school I couldn't get enough information about agriculture. The journey is as much fun as getting there', stressing that enjoyment is high on Margrave's agenda. He also believes: 'I'm quite creative with a good imagination – a critical skill. You must imagine and write down your thoughts....'

And certainly Hank writes a lot. He makes lists of the jobs that must be accomplished in priority order, and constantly reviews both what to keep on the list, and in what order. Hank is also a firm believer that experience is the key to success, or rather, as he puts it 'technical knowledge + natural ability + experience gets the right jobs done'. He adds 'to interpret experience you need experience to understand why'. Hank's list, then, includes:

- the ability to mull over experiences and learn the lessons on offer;
- an ability to read a lot and absorb technical information for general use;
- good labour management skills;
- good skills at using spreadsheets and doing the sums on alternatives;
- an ability to judge the outcomes from changing the farming system; and
- the confidence in your conclusions to act decisively.

Hank believes he learnt many lessons on managing labour from early mistakes. Initially Hank tried to direct his helpers with detailed instructions, but soon learnt many resented such specific control. Hank learnt you had to 'give employees scope'. It is interesting that Margrave quickly came to the same conclusion which ended up with him giving each person an area of responsibility.

With the volatility of prices and conditions, Hank believes it is important to constantly update cash flows and budgets so you are aware of the changing bottom line and bank overdraft. Quick decisions can then be made about problem situations. It is no use shutting the gate after the horse has bolted. This belief is reflected in his list of skills, and certainly with developing a new farm involving vast investments good control of the cash flow situation is imperative.

During his time in a rural bank Hank learnt to quickly assess a proposal and decide if, and how much, the bank would lend. This experience has stood him in good stead. Hank notes the toughest decision he has faced in recent times is which milk processing company to contract to for his new farm. Corporations are not always totally forthcoming with their information, for, no doubt, it helps to keep some of the details somewhat hidden. In the end, after doing many long-term budgets and world market analyses, Hank decided to go with the cooperative with which he was familiar. Perhaps this reflects his risk-averse nature. He concluded 'that was a real hard decision'.

Introducing Two Consultants, and Their Views on Skills

Bruce and his dairy farming interest

Bruce really wanted to be a farmer, but his access to the necessary resources precluded this option when he was young, and now that purchasing a farm would be an option, Bruce believes he would find some of the physical jobs required rather boring, but he does enjoy the decision-making side of agricultural production. This is one of the reasons he is pleased he followed the path into consultancy. But to satisfy his almost inherent desire to be involved in practical farming he has a small part-time farm.

Bruce was brought up in a city, though his family had country friends where he spent as much time as possible. While his father was a city businessperson with little interest in agricultural production, Bruce believes some of the urge to be a farmer comes from his farming ancestors in Scotland. This agricultural fascination led Bruce to take agriculture at university where again most of his friends were from farms, so holidays saw him visiting one farm or another. His best friend was from a dairy farm so perhaps these experiences sparked the keenness to be involved in this industry. Bruce even emulated farmers by wearing typical farmer attire.

Bruce believes he is risk-averse as otherwise he might just have scraped together what funds he could find and started land acquisition. Instead the university path was followed, and eventually postgraduate study. A strong involvement in the Boy Scout movement had Bruce taking many leadership courses and ending up as one of the leaders. This training, and perhaps his father's interests, led Bruce into management and agricultural economics courses to satisfy this fascination with organizing, management and decision making. As Bruce comments: 'I was always intrigued with how to put resources together and make money'. This interest also stemmed from his romantic visions of being a farmer, but the family and friends reckoned a city boy could never join the farming community. They were wrong.

Bruce left university and went straight into consulting positions, and on reflection wonders how he managed so well given his lack of experience particularly after making an initial mistake which knocked his confidence. However, he did overcome problems, and successfully related to farmers, becoming, in the end, a strongly respected consultant dealing with both individual farmers as well as groups over technical issues in dairy farming.

When giving his views on the skills successful farmers require, Bruce comments that farming is not something which you pick up overnight. To Bruce, as a newcomer that has taken a long time to absorb the unspoken nuances, this is an important observation. Bruce noted in talking about farm family dynasties 'succession is not just the land, but all the skills and heritage that get absorbed almost by osmosis'.

'There is heaps of stress in agriculture' is an observation that leads Bruce into noting an important skill is handling risk. 'A crop man invests thousands into cultivation, seed, and fertilizer in the hope that it will rain'. Who else would have such confidence? This is one reason why good farmers 'must not be scared of debt...for without borrowing little progress can be made'.

A farmer must also be an assiduous observer noting all changes, and quickly respond appropriately. 'You can't just stay static'. And in this process success requires the farmer to have excellent decision-making skills with an ability to judge the information being received, analyse it and act accordingly. In a nutshell, the farmer must be good at responding to change.

Bruce also believes farmers must be totally open-minded and assess each situation on its merits for improving their skill is dependant on learning from mistakes. Thus, an important skill is being able to say 'hey, I've made a mistake, how can I learn from this?'

This 'willingness to learn creates a good manager'. And 'if the farmer exposes himself to stress some learn and improve, others sink'. This is all part of 'being competitive and resilient' and not wanting 'other farmers to beat me'. Bruce comments that the 'personal factor' is vital to good management. By this he means the farmer's honesty, integrity and work ethic: 'Can he meet a challenge, and go the extra mile?'

Thus, a farmer's competitive nature is an important attribute in driving him to improve, 'do they want to get respect, or just plod on?' Bruce comments that a farmer's entrepreneurial skill is part and parcel of success. If you expose them to possibilities, 'some pick them up, others don't. They need this innate ability to start with. It is very difficult to help someone without this desire'.

A farmer must have strong prioritization skills. Bruce talks about a quadrant of jobs with each being given a grading based on how important they are, and the urgency with which they must be attended to. Thus, you have non-important jobs with little urgency through to very important jobs that require to be completed with urgency. It is only this latter grouping that should be attended to. Hank the dairy farmer clearly has this prioritization skill using his constantly updated diary to sort out the jobs in this 'to do' quadrant.

No doubt given more time Bruce could have added to his list of skills, but the list discussed is what stands out in his mind as being associated with high managerial skill.

'Prof' and his comments about the skills required

The second consultant we will call 'Prof' largely as he has been involved in a wide range of activities within agriculture including working as an academic and researcher. This wide range of experience means he has come to well-based conclusions on farmers and their modus operandi through many different eyes. Prof started his professional career as a 'farm adviser', as they were called then, but today he would have been called a consultant. This career then moved into research, and, finally, tertiary teaching and research student supervision. Through the years Prof has also been involved with international consulting, some of which was helping tertiary establishments develop farm management teaching. Prof has also supervised a research farm that attracted hundreds of farmers interested in the ground-breaking systems developed. In a nutshell, Prof is eminently qualified to comment on what makes a successful farmer.

You might imagine Prof was stimulated to become involved in agriculture through his early experiences and contacts, but nothing could be further from the truth. Prof was a city boy who had no contacts at all in rural areas, and his relatives and recent ancestors have all been urban people. One wonders where the strong desire to be involved in primary production came from. As Prof notes: 'I just had this strong interest even though I went to a high school that was only interested in the traditional professions... engineers, doctors, lawyers, accountants...'. Prof's father was an engineer, who was very interested in his son's activities, development and career, and whose only stipulation was that Prof should go to university (Prof's first university was agriculturally based. See Fig. 5.3) rather than work on farms. In fact, Prof's first contact with farms was spending 3 days with a farm adviser to gain insight into what was involved. He organized this contact himself while still at school, so there was a strong motiva-

Fig. 5.3. This is where Prof's journey into agriculture started.

tion stemming from some inner source. Prof's first practical experience of agriculture was spending the summer holidays immediately before moving to university on a dairy farm. This dairy interest has never left him, and he still takes a strong interest in the development of the industry.

The other great love Prof immersed himself in was mountaineering. While still at school he joined a tramping club and enjoyed their activities noting that 'I had this great attraction to the wide outdoors', and perhaps this was part of the desire to be in agriculture. The mountaineering love led Prof to take time off work and education to visit a wide range of mountain ranges including high peaks in South America, the Himalayas and Antarctica. The challenges and demands of high mountaineering clearly developed Prof's determination, survival abilities and people skills which, in turn, helped his professional career. Prof comments: 'one of the reasons for doing post graduate study was the longer holidays which enabled following my passion for the mountains'.

Today Prof leads a very busy life teaching, writing, addressing different groups including farmers, and international consulting. However, this schedule now precludes spending much time in the mountains. The price of success is sometimes hard to take, but through all this, Prof has clear views on the skills a farmer should have if he is to be similarly successful.

Top of the list is technological knowledge and expertise. 'This goes without saying', he comments, for it is the foundation from which success occurs provided successful application is achievable. The main other key is being able to put the technology into workable systems. The farm is an integrated whole of many connected components so it is important to create from the jigsaw parts something that achieves the objectives. A strong curiosity is one of the skills that enables the successful systems to be put together, a strong curiosity to search out new ideas and ways of integrating them into a whole. Prof believes a farmer faces one of the most complex decision set-ups existing in modern society, and creating a successful system is challenging, but something that farmers in general do well. In this respect Prof holds farmers in extremely high regard.

An important skill, according to Prof, is the ability to observe what other farmers are doing and consequently weeding out the better systems which can then be applied 'at home'. Thus, networks are important, and benchmarks which the farmer has in mind so that he can assess the possibilities and weigh them up against his current system. An ability to create a good team about them is an important skill enabling an assessment of all the ideas and their integration into a good total system. While the farmer might not have all the skills themselves, Prof notes that 'the ability to know your strengths and weaknesses leads to selecting a team that together forms a complimentary package'.

Agreeing with the other farmers and consultant, Prof stresses that in agriculture the successful manager must be passionate about agriculture. If he is not, he just won't succeed, for farming is nothing like an 8 to 5 job after which you move on to other parts of your life. Prof comments that 'a good farmer will be constantly mulling over all the options, strategies and tactics no matter where he is in an effort to come up with the best decisions and actions. This requires dedication, total commitment, and this passion for being involved in agriculture'.

One of the core reasons for the need for constant thinking is the risk and uncertainty that surrounds agriculture. 'You can't make a decision then walk away from it for something is bound to change requiring a re-evaluation'. Thus another important skill is the ability to be flexible. As Prof puts it 'you need to get yourself into a space where opportunities arise as conditions change'. This is partly a frame of mind as well as having a range of options at your fingertips. However, Prof comments, 'this doesn't mean you don't plan. In fact planning is everything as going through the planning process puts you in a position to take the opportunities'. You need to 'think through the strategies and tactical responses for a farmer works in a complex uncertain environment'. However, the best approach varies with the type of farming. Some production systems are akin to a factory, examples are pig and poultry production as well as feed lots. In these situations, production is largely isolated from the weather so an important skill is being able to plan systems over long periods of time.

Good management requires the farmer to be self-aware and critical. Prof notes that a farmer must examine all actions with a strong yardstick using sensible benchmarks. In other words, complacency is an enemy, and 'the arrogant farmer won't make much progress'. Thus, an open mind is an important attribute preventing preconceptions from blocking a proper examination of alternatives and opportunities.

As noted, Prof is a strong believer that a farmer should be fully aware of his personality so that he understands his strengths and weaknesses. Prof recounts the use of standard tests to enable members of a management team, often a husband and wife, to examine how they fit together and, therefore, where they may need to make special efforts to support and control behaviour. Prof also talks about the use of the Margerison–McCann Team Management Wheel (www.tms.com.au/tms07.html), in which people are classified into eight types. These are the reporter–adviser, creator–innovator, explorer–promoter, assessor–developer, thruster–organizer, concluder-producer, controller–inspector and upholder–maintainer. The right mix can enhance a team, but even if a good team does not fall out given the people involved, at least an understanding of each type helps people adjust.

In a nutshell, Prof believes the following skills and attributes are all important to success in farm management:

- an excellent technological knowledge and ability to implement the chosen systems;
- an ability to develop integrated whole farm systems;
- having a strong curiosity;
- successful at observing what other farmers are doing, and judging what is observed using benchmarks;
- an ability to create a team that compliments each other;
- knowing and understanding your own personality, strengths and weaknesses;
- having a strong passion for agriculture;
- an ability to constantly review outcomes, options and alternatives;
- an ability to be flexible in contrast to preferring to work with fixed systems; and
- self-awareness and an open mind.

Can you add to this list? Prof also talked about the need for labour management skills now that many dairy farms in particular are employing as many as 12 people. Some farmers find this number difficult to handle. One farmer Prof talks about found his personality just did not fit in with large numbers of employees so he cut back his operation to five people as he could then work his favoured personal mentoring system with each labour unit. Each was much happier and worked well together.

Concluding Comments

There are major differences between farmers in their managerial skill. One wonders why these differences occur, and what can be done to improve the skills of all farmers, good and bad. It is probably true that most skills are learnt from experience, often before becoming a manager, especially where the manager has lived on a farm from a young age. Thus, most farmers grow into the management role and will have a full range of skills, whether good or not so good, through this process. However, this chapter has highlighted the core skills that all farmers will have even if they cannot list them out themselves. This list enables a systematic consideration of the components of a farmer's total skill, and which sub-skills might need extra attention. Similarly, the list is a blueprint for the skills that need including in any formal courses on improving management.

The ranking of the skills was obtained from a large group of sophisticated and well-educated farmers involved in businesses valued at around US$1,100,000 on average. However, it would be surprising if the skill list would be much different in less-developed farming situations. One area that might be more important is 'people skills' due to the wider involvement of family and community as shown by studies quantifying decision processes in less developed countries.

It is also important to note that the type of farming did not seem to impact on the core skill list, though this would be expected as details of, for example, the particular technology knowledge required was not investigated, nor the specific technical skills.

Finally, it is worth noting that one of the highly ranked specific skills was 'having an adventurous spirit'. In general, primary production is certainly other than a controlled factory situation where processes and procedures are standard fare. The challenges and complexities are such that people with an adventurous spirit are more likely to succeed.

Appendix 5A. Survey on Managerial Factors

Please complete and return this questionnaire using the enclosed postage-paid envelope. All information provided will be kept in strictest confidence to the researchers involved. If you are not the operator or manager of the property please pass the questionnaire on to this person.

A. GENERAL

1. **Farm type.** Please tick ONE box representing the MAJOR enterprise type on the property you operate.

    Intensive sheep   ❑   Extensive sheep ❑   Deer ❑   Cattle   ❑
    Dairying   ❑   Other animal   ❑   Fruit ❑   Cash crop ❑
    Ornamental/flowers ❑   Vegetable   ❑   Other ❑

2. **Labour.** Including the manager, please give the number of equivalent full-time adult people it takes to run the property (use fractions if necessary, e.g. three-fourths).

3. **Area.** What is the total land area used in the operation, including rental/leased land? acres/hectares (Cross out the acres or hectares sign depending on the unit used)

B. IMPORTANCE OF MANAGERIAL ATTRIBUTES

For your situation, please rate the importance of each of the managerial attributes listed below.

Use a score range of **7** (VERY important) to **1** (NOT AT ALL important) with **4** representing MODERATELY important and the other numbers for in-between degrees of importance.

1. Ability to identify the key factors in a problem and discard the irrelevant. _____ ☐

2. Quickly analysing and sorting out situations that have never been faced before. _____ ☐

3. Having a clear understanding of the family's objectives, values and goals, thus making assessing the value of alternative actions easy. _____ ☐

4. Being able to predict local weather better than the official forecaster. _____ ☐

5. Understanding the local political scene as it might impact on rules affecting what can be done. _____ ☐

6. Developing and maintaining a support network of colleagues and professionals. _____ ☐

7. Being able to efficiently organize and carry out quite complex operations (e.g. get a new packing shed operational on time). _____ ☐

8. Developing appropriate and detailed plans for both short- and longer-term horizons. _____ ☐

9. Making requirements clearly understood (effective communication). _____ ☐

10. Understanding the basis on which to choose between alternatives (e.g. knowing how to cost unpriced labour, knowing how to do gross margins, understanding diversification principles). _____ ☐

11. Being up-to-date with the current condition of the property in its totality (bank balances, animal condition, crop growth, soil moisture, feed levels, machinery repair, etc.). _____ ☐

12. Picturing (understanding) the consequences of a decision over the many (or few) months/ years it might impact over (e.g. planting an area in forestry, subdividing a paddock, etc.). _____ ☐

13. Skill at keeping, interpreting and using recorded data about the property and associated factors (e.g. market trends). _____ ☐

14. The ability to predict product prices into the foreseeable future, or at least understanding the factors that determine the prices, and understand market requirements. _____ ☐

15. Assessing job priorities. _____ ☐

16. OTHER – If you think an important managerial component has been left off the list, please write it below and give it a score. _____ ☐

    (i) _____ ☐

    (ii) _____ ☐

C. IMPORTANCE OF ENTREPRENEURIAL SKILLS

For your situation, please rate the importance of each of the entrepreneurial skills listed below.

Use a score range of **7** (VERY important) to **1** (NOT AT ALL important) with **4** representing MODERATELY important and the other numbers for in-between degrees of importance.

1. Being able to seek out, identify, and clarify new opportunities (production, products, marketing, etc.)._____

2. An ability and determination to look/ask/seek out information thought to be necessary for making decisions. _____

3. Ability in learning new skills._____

4. An intuition that gives early warning signs when something is right, or, in contrast, when something positive needs exploiting. _____

5. Skills in finding the very best market (price, quantity, etc.) for all output._____

6. Understanding deadlines and being able to 'act in time' (e.g. spray before insect damage, fertilizer applied in good time)._____

7. The skill to negotiate the best possible deal (price, arrangement, etc.)._____

8. A belief in being able to control a lot of what happens around the property in contrast to a belief that not much is really controllable due to the weather, markets, government action, etc._____

9. When faced with opportunities, ensuring ALL alternatives are sought out, considered and evaluated._____

10. The skill and intuition to forecast well into the future likely opportunities in products and production systems._____

11. An ability to look ahead and anticipate likely problems, needs, and opportunities._____

12. Understanding sources of risk and what can be done to reduce its impact._____

13. OTHER – If you think an important entrepreneurial component has been left off the list, please write it below and give it a score.

    (i) _____

    (ii) _____

D. MANAGERIAL STYLE

For each of the following statements indicate how true it is with respect to your management style. Each question has five boxes beside it – tick only the ONE that best records the degree of truth in the statement.

1. You tend to mull over decisions before acting. TRUE ❑ ❑ ❑ ❑ ❑ NOT TRUE

2. You find it easy to ring up strangers to find out technical information. TRUE ❑ ❑ ❑ ❑ ❑ NOT TRUE

3. For most things you seek the views of many people before making changes to your operations. TRUE ❑ ❑ ❑ ❑ ❑ NOT TRUE

4. You usually find discussing everything with members of your family and/or colleagues very helpful. TRUE ❑ ❑ ❑ ❑ ❑ NOT TRUE

5. Where there are too many jobs for the time available you sometimes become quite anxious. TRUE ❑ ❑ ❑ ❑ ❑ NOT TRUE

6. You tend to tolerate mistakes and accidents that occur with employees and/or contractors. TRUE ❑ ❑ ❑ ❑ ❑ NOT TRUE

7. You share your successes and failures with neighbours. TRUE ❑ ❑ ❑ ❑ ❑ NOT TRUE

8. Keeping records on just about everything is very important. TRUE ❑ ❑ ❑ ❑ ❑ NOT TRUE

9. You admire farming/grower colleagues who are financially logical and don't let emotions colour their decisions. TRUE ❑ ❑ ❑ ❑ ❑ NOT TRUE

10. You sometimes don't sleep at night worrying about decisions made. TRUE ❑ ❑ ❑ ❑ ❑ NOT TRUE

11. You find investigating new farming/growing methods exhilarating and challenging. TRUE ❑ ❑ ❑ ❑ ❑ NOT TRUE

12. You tend to write down options and calculate monetary consequences before deciding. TRUE ❑ ❑ ❑ ❑ ❑ NOT TRUE

13. You tend to worry about what others think of your methods. TRUE ❑ ❑ ❑ ❑ ❑ NOT TRUE

14. You are happy to make do with what materials you have to hand. TRUE ❑ ❑ ❑ ❑ ❑ NOT TRUE

15. You find talking to others about farming/growing ideas stimulates and excites you as well as increasing your enthusiasm for new ideas. TRUE ❑ ❑ ❑ ❑ ❑ NOT TRUE

16. Having to make changes to well-established management systems and rules is a real pain. TRUE ❑ ❑ ❑ ❑ ❑ NOT TRUE

17. You normally don't rest until the job is
fully completed.   TRUE ❏ ❏ ❏ ❏ ❏ NOT TRUE

18. You normally enjoy being involved in
farmer/grower organizations.   TRUE ❏ ❏ ❏ ❏ ❏ NOT TRUE

19. You sometimes believe you are too much of
a stickler for checking and double-checking
that everything has been carried
out satisfactorily.   TRUE ❏ ❏ ❏ ❏ ❏ NOT TRUE

20. When the pressure is on you sometimes
become cross and short with others. TRUE ❏ ❏ ❏ ❏ ❏ NOT TRUE

21. You generally choose conclusions from
experience rather than from hunches when
they are in conflict.   TRUE ❏ ❏ ❏ ❏ ❏ NOT TRUE

22. You are inclined to let employees or
contractors do it their way.   TRUE ❏ ❏ ❏ ❏ ❏ NOT TRUE

23. You not only speak your mind and ask
questions at farmer/grower meetings,
but also enjoy the involvement.   TRUE ❏ ❏ ❏ ❏ ❏ NOT TRUE

24. It is very important to stick to management
principles no matter what the pressure to
do otherwise.   TRUE ❏ ❏ ❏ ❏ ❏ NOT TRUE

25. You are much happier if everything is
planned well ahead of time.   TRUE ❏ ❏ ❏ ❏ ❏ NOT TRUE

E. GOALS AND AIMS

For each of the following statements indicate how true it is with respect to your goals and aims. Each question has five boxes beside it – tick only the ONE that best records your degree of belief in the statement.

1. It is very important to pass on the property
to family members.   TRUE ❏ ❏ ❏ ❏ ❏ NOT TRUE

2. It is important to earn the respect of
farmers/growers in the local community. TRUE ❏ ❏ ❏ ❏ ❏ NOT TRUE

3. Making a comfortable living is important. TRUE ❏ ❏ ❏ ❏ ❏ NOT TRUE

4. It is very necessary to keep debt as low
as possible.   TRUE ❏ ❏ ❏ ❏ ❏ NOT TRUE

5. It is essential to plan for reasonable holidays
and plenty of leisure time.   TRUE ❏ ❏ ❏ ❏ ❏ NOT TRUE

6. Attending field days and farmer/growers
meetings is vital.   TRUE ❏ ❏ ❏ ❏ ❏ NOT TRUE

7. It is very important to reduce risk using
techniques like diversification, farming
conservatively, keeping cash reserves. TRUE ❏ ❏ ❏ ❏ ❏ NOT TRUE

8. Developing facilities and systems that give good working conditions is crucial. TRUE ❑ ❑ ❑ ❑ ❑ NOT TRUE

9. It is very important to ensure employees enjoy their jobs. TRUE ❑ ❑ ❑ ❑ ❑ NOT TRUE

10. Doing jobs that I enjoy is a very important part of the operation. TRUE ❑ ❑ ❑ ❑ ❑ NOT TRUE

11. Minimizing pollution is very important. TRUE ❑ ❑ ❑ ❑ ❑ NOT TRUE

12. I enjoy experimenting with new products and production systems. TRUE ❑ ❑ ❑ ❑ ❑ NOT TRUE

13. Proper retirement planning is a major consideration. TRUE ❑ ❑ ❑ ❑ ❑ NOT TRUE

14. You must always be striving to increase the total value of assets. TRUE ❑ ❑ ❑ ❑ ❑ NOT TRUE

15. Constantly expanding the size of the business is absolutely necessary. TRUE ❑ ❑ ❑ ❑ ❑ NOT TRUE

16. Aiming for maximum sustainable net cash returns is very important. TRUE ❑ ❑ ❑ ❑ ❑ NOT TRUE

17. Maintaining a presence in local community activities is important. TRUE ❑ ❑ ❑ ❑ ❑ NOT TRUE

18. It is very important to improve the condition of the property (fertility, facilities, etc.). TRUE ❑ ❑ ❑ ❑ ❑ NOT TRUE

19. Giving assets to the children so they can pay for education and/or set up businesses is very important. TRUE ❑ ❑ ❑ ❑ ❑ NOT TRUE

F. COMPUTER USE

If a computer is used for business on your property, give the average HOURS PER MONTH that it is used for the following (otherwise go to the next question).

Recording financial transaction information ⎯⎯⎯⎯   ▭

Doing forecast budgets/cash flows ⎯⎯⎯⎯   ▭

Keeping animal records ⎯⎯⎯⎯   ▭

Keeping paddock/product records ⎯⎯⎯⎯   ▭

Word processing ⎯⎯⎯⎯   ▭

Searching the Web for information ⎯⎯⎯⎯   ▭

Sending e-mails ⎯⎯⎯⎯   ▭

Entertainment/education ⎯⎯⎯⎯   ▭

Internet banking ⎯⎯⎯⎯   ▭

Internet purchasing ⎯⎯⎯⎯   ▭

OTHER ⎯⎯⎯⎯   ▭

G. IMPORTANCE OF PERSONAL ATTRIBUTES

For your situation, please rate the importance of each of the personal attributes listed below.

Use a score range of **7** (VERY important) to **1** (NOT AT ALL important) with **4** representing MODERATELY important and the other numbers for in-between degrees of importance.

1. Early observation of important indicators around the property (e.g. lambs are scouring, wheat is infected, cows losing weight, pasture growth has increased, and so on). ＿＿＿＿＿

2. Keeping a cool head and putting aside any tendency to panic when faced with stressful situations. ＿＿＿＿＿

3. Having the confidence to draw conclusions and act quickly and decisively. ＿＿＿＿＿

4. An excellent knowledge of facts, figures, procedures and methods, with respect to soils, plants, animals, machines, buildings, etc. ＿＿＿＿＿

5. Being prepared to give it a go and take risks in changing production systems and/or starting new ventures. ＿＿＿＿＿

6. High motivation in constantly seeking better ways and implementing them; in contrast to being happy with current systems. ＿＿＿＿＿

7. Accepting the good and the bad, and not letting it affect management and decision making. ＿＿＿＿＿

8. Ability to learn from experience, mistakes, and failures. ＿＿＿＿＿

9. The determination to keep working all hours until the high-priority jobs are completed. ＿＿＿＿＿

10. Developing a 'good moral character' involving openness, integrity, reliability, trustworthiness, etc. ＿＿＿＿＿

11. Developing a strong personality so that others' 'sit up, notice, respect, and act' on what is said. ＿＿＿＿＿

12. Understanding the interrelationships between all the components of the property (e.g. rainfall–soil moisture–plant growth–animal grazing, i.e. what affects what?). ＿＿＿＿＿

13. Obtaining employees' and/or contractors' cooperation and understanding leading to harmonious and productive relationships. ＿＿＿＿＿

14. Tertiary education in areas related to primary production (agriculture, horticulture, biology, marketing, etc.). ＿＿＿＿＿

15. Having above average intelligence and school grades. ＿＿＿＿＿

16. Successfully judging personality and selecting suitable employees.＿＿＿＿＿

17. Successfully resolving conflicts on, and off, the property (e.g. dispute between employees). ⸻ ☐

18. Maintaining good relationships with outside people – bankers, accountants, suppliers. ⸻ ☐

19. OTHER – If you think an important personal attribute has been left off the list, please write it below and give it a score. ⸻ ☐

   (i) ⸻ ☐

   (ii) ⸻ ☐

H. PERSONAL FEATURES

1. Which age group do you fall into?   (Tick ONE box)

   Less than 25 years ☐   26–35 years ☐   36–45 years   ☐
   46–55 years   ☐   56–65 years ☐   Greater than 65 years ☐

2. What was the level at which you stopped your formal education? (Tick ONE box)

   Primary school   ☐   Secondary school – up to 3 years ☐
   Secondary school – 4 or   Tertiary education – up to 2 years ☐
   more years   ☐
   Tertiary education– 3 or
   more years   ☐

3. Please indicate your gender by putting **F**(emale) or **M**(ale) in the box.   ☐

4. Please rate yourself in general intelligence – tick ONE box. (If you are uncomfortable answering this question, leave blank)

   Highly intelligent ☐   Reasonably intelligent ☐   Average
   A bit below average ☐   Other ☐   intelligence ☐

5. If all farmers were rated on a **10** (excellent) to **1** (poor) scale for managerial ability, what level of skill rating would you give yourself?   ☐

MANAGERIAL TRAINING

1. To what degree would you use a managerial skill training programme, if available in your area?   (Tick ONE box)

   Not at all ☐   Occasionally ☐   Extensively ☐

2. Assuming training was available, please rank the following method of delivery in order of preference (**1** for most preferred, **2** for the second preferred, and so on).

   Computerized self-training   ☐   Book-based self-training ☐
   Tutored system based locally ☐
   OTHER (please specify) ⸻

3. On what topics/skills would you like training?

   (i) ⸻ ☐

   (ii) ⸻ ☐

   (iii) ⸻ ☐

6 Biases and Stress

Introduction

The majority of farmers do not get it right on all occasions. Mistakes are made in one or more of the attributes or skills that are involved in a good decision. These skills were listed in the last chapter (Chapter 5). If, however, a farmer is consistently wrong in some of the skills, then his decision making can be said to be *biased*. For improvement, the bias must be recognized, and the error of his ways corrected, so that in future mistakes are only random in contrast to consistent. This chapter contains a description of the more common biases, and provides comments on their recognition and correction. Of course, in many respects the full list of potential biases has already been given in that a consistent aberration in any of the skills required is a bias. However, it is still useful to highlight the common biases that researchers have particularly noted.

Then there is the problem of stress. Primary production involves considerable risk and uncertainty and, consequently, gives rise to considerable stress. There is nothing more disheartening than planning for developments, committing a lot of money, only to find that the project fails due, say, to a particularly poor year climate-wise. The feeling of a lack of control can be overpowering. It is not uncommon to see reports of farmer suicides under extreme situations. Stress, then, is common, but what matters is how it is handled and whether it impacts on the effectiveness of a farmer's skills. A section of the chapter covers types of stress and what might be done to help remove the impacts of stress.

It is not yet exactly clear what creates a bias in a farmer, and why a farmer has excessive stress, but clearly the farmer's personality, intelligence, family background and experience must all impinge on the creation of biases, and stress. The education system must also give rise to some of the biases. No one wants to acquire biases, nor overbearing stress, and so in most cases their existence is due to oversights of various kinds. It must be remembered that everyone makes mistakes from time to time, particularly due to chance elements (it is always easy to see what should have happened in hindsight, but no one has a perfect prediction record), so care must be taken in deciding whether a bias exists, or whether, on average, the correct approach is being used. Similarly, some stress is normal, and indeed can lead to careful and correct decision making, but if the stress is too much, decision efficiency declines, and certainly enjoyment also declines similarly.

Biases usually relate to decision making in contrast to the physical activities around a farm. Clearly, however, physical prowess, or the lack of it in a consistent sense, will cause problems. Thus, someone who thinks the method they use when, say, vaccinating animals is appropriate, when in fact it is spreading an infection has a 'biased' procedure. These physical biases are not discussed here.

To categorize potential biases it is useful to think of the decision steps and biases in each step. Figure 6.1 lists the steps. The diagram reflects that information and data exist to be observed and noted, similarly the world around contains markets of various kinds which determine the factors which pass to and from the farm entity, for example, prices of the products. Finally, the interaction of inputs and outputs is governed by a set of biological relationships such as animal output relative to feed input. The farmer observes all this and decides what the values of the important information and data are, including market information. Whether these observations provide the relevant variables, and an accurate estimate and understanding of them, is another matter. This is reflected in the next row of boxes. The brain held information is then exposed to the farmer's processing to finally come up with a conclusion leading to action and outcome.

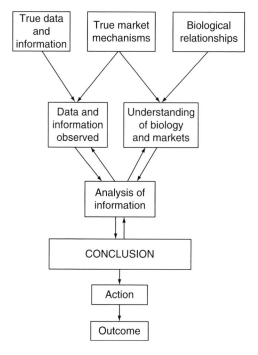

Fig. 6.1. Decision process sources of bias.

There are also feedback loops in that, for example, if the conclusion reached appears unusual, the farmer may re-sample the information to come up with, say, a better estimate of the product price. Similarly, following the observation of the outcome, various assumptions and information might be altered.

Considering this whole process, biases can be grouped according to the boxes in Fig. 6.1. Thus, we have:

- observational biases;
- forecasting biases;
- decision process biases;
- implementation biases; and
- general biases that might impact on several areas.

In addition, as a farmer works with people in the process of creating decisions, and their implementation, another bias category can be called:

- people dealing bias.

This allows for inappropriate relationships that lead to consistently less than optimal outcomes. Strictly this bias could be put under the 'implementation' label in part, and also under the 'observation' label.

The discussion that follows will consider each of the bias areas in turn, and conclude with comments about stress before finally covering suggestions on observing and correcting biases remembering that a bias has been defined as an error which is made consistently, and the decision maker, usually, is unaware of its existence.

Areas of Bias

Observation

Most farmers rely on their memory for a significant proportion of their decision information and so accuracy depends very much on picking up the true data as they are observed, and then storing it in such a way that they are correct when retrieved. The trouble is that memory imperfections (bias) are common. For example, the context in which something is observed effects what is stored and remembered. If, for example, a farmer attends a meeting at another farm and is shown an obviously excellent crop together with the yield data that have resulted from a particular fertilizer application, it is likely this information will be easily retrievable. In contrast, if the farmer reads about the same situation in a back corner of a rural newspaper, or hears a radio report, the information is less likely to be observed for successful future retrieval. In that most farmers are what can be called 'concrete', or kinaesthetic, learners so that seeing and doing tends to be a more successful way of observing. Thus, a *farmer's learning approach can lead to biases*.

Then there is the *halo effect*. If something good is observed about an item, it is often assumed that all other characteristics of the item are also good. For example, if a crop of wheat has a high yield, the halo effect has the farmer also assuming that the quality of the wheat is good. Clearly, these two parameters may well be divergent.

Another common problem is called the *anchoring effect*. This occurs where the introductory information provided influences the conclusion the farmer makes on the value of a variable. For example, it has been found that people will come to different conclusions on the monetary value of an asset to be sold depending on the preliminary information provided or observed. So, if you said a farm had sold for US$1,000,000, 3 years ago, you often get a different estimate of today's value if in contrast, say, you told the valuer that the farm sold 4 years ago for US$900,000. A farmer must be aware of this common bias, and adjust accordingly.

The *framing impact* is similar to the 'anchoring' effect. This refers to the impact of the information, or deliverer, surrounding the core information. Thus, a farmer might come to a different conclusion, if, say, his spouse reads out information compared to the same information being read out by a professional consultant. The same sort of impact can occur if the material is presented in a professional glossy form relative to, say, a simple note even though the core information is identical.

Selective abstraction (or sometimes called 'selective perception') is another common bias. This refers to remembering only the data that suit you (for whatever reason). Thus, for example, a farmer might only remember the yields of a crop for favourable seasons, so that when budgeting expected income the 'average' yield used is much higher than the true average. This is something that most of us do in that you remember the good things with the bad being shut out. Perhaps this is a survival attribute in that people tend to push the bad to the back of their mind so that they are positive about the future, whereas a person who has had a long-run of bad luck might start to wonder if it is all worthwhile. Similarly, how often have you selected information that suits your feelings and so ignore the data that perhaps lead to a different conclusion? Thus, for example, if a farmer really enjoys dealing with stock he may well put the lowest possible crop yields in his budgets when comparing crop profitability with animal returns. The conclusion, clearly, reinforces his personal wants. This might not be a bad thing provided it is recognized that excluding crops is in fact foregoing profit and this is a cost of sticking to animal farming.

The old saying says 'one swallow does not make a spring', yet one good outcome often provides a farmer with the conclusion that good outcomes will always occur. This is referred to as the *sample-size* effect. For objectivity a farmer should adjust his estimates of yields, outputs and costs according to the sample size on which the estimate is based. This bias is particularly relevant for farmers starting in a new area.

Finally, while the farmer may not have any reason to make biased estimates, he may simply be using a *flawed procedure*, or is insufficiently trained to get the estimate correct, or reasonably so. For example, many farmers will use an eye estimate of the amount of pasture, or a crop, available for grazing. Similarly they may use an eye estimate to judge the live weight of animals. Many such examples exist. However, the farmer may be consistently underestimating or overestimating the true values. Experiments have shown that with proper training a farmer can be both accurate and consistent in estimating all the examples given. Often this training requires tests and feedback until accuracy occurs. In the case of pasture, this would involve making an estimate, then cutting a sample and, after drying, weighing it. This is repeated from time to time until constant accuracy occurs. Similar procedures would be used for animal live weight estimates.

It must also be remembered that the farmer must 'observe' the production unit's objectives correctly. The objectives, of course, serve as the measuring stick when deciding which alternative is to be followed. This assessment must account for the farm family's objectives as well as those of any other owners in the farm. Not all farmers, of course, see it this way in that they believe it is their own personal objectives which should take precedence. This situation is fraught with potential biases; thus, *bias in concluding on the appropriate objectives* must be guarded against.

In summary, in the general area of observation, recognized biases are:

- flawed learning approach;
- the 'halo effect';
- the 'anchoring effect';
- the 'framing impact';
- selective abstraction;
- sample-size effect;
- flawed procedures; and
- misleading objectives.

No doubt, others also exist.

Forecasting

Having rightly or wrongly observed data and information, this must be turned into useful managerial material. This invariably involves forecasting future situations resulting from one or more decisions. Thus, a farmer needs to forecast, say, the yield of maize resulting from using a particular cultivar, a particular fertilizer, irrigation, weed control, disease and pest control regimes in order to assess the crop profit distribution, and therefore whether to grow the crop relative to alternatives. This estimate also requires the market and price situation to be forecast. No matter which part of a farm is considered, decision making involves forecasting in some form. Failure to get this correct on a consistent basis gives rise to bias and incorrect decisions.

Common biases in this area include:

- 'dichotomous' thinking;
- technology misunderstandings;
- risk estimate failures;
- giving credence to un-evidenced guesses;
- 'availability and recency' effects; and
- generalization effects.

Dichotomous thinking involves a farmer assuming that an outcome will be either good or bad. In reality most outcomes in the uncertain world of farming follow a continuous curve from the good extreme to the bad extreme. In assessing an option, just taking the good and the bad possibilities may distort the result for the bad may overbear the good, whereas there will be many possible 'good' outcomes that may produce the opposite impact.

The main reason for following the dichotomous approach is its ease of estimation. It is possible to mentally calculate the result of both a good and bad outcome and weigh them up. To do more than this requires pencil and paper using chance (probability) estimates.

Once you start thinking about chances, or probabilities, a farmer's understanding of risk and risk calculations comes into the picture. Farmers tend to produce imperfect estimates of the probability of the various possible outcomes (*risk estimate failures*). Probability distributions are a further complication to the farmer and are seldom allowed for in the mental, or otherwise, sums. The notion of an 'expected outcome' (the sum of the possible outcomes each weighted by its probability of occurrence) is difficult for farmers to grasp. They often think in terms of single outcomes as this is less challenging than the idea of a distribution from which only one event will occur in any one year or trial. How do you accept the idea that the decision made turns out to be incorrect in hindsight where in reality it was indeed the correct decision, it is just that chance created one of the less likely set of circumstances. Perfect prediction is not possible in a random world, well not yet anyway. The overall outcome is that farmers do tend to struggle with risk and uncertainty and tend to work with single-valued estimates; these might be called 'certainty equivalents'.

If a farmer is unaware of the *true technical relationships* determining outcomes from an input set, then clearly the forecasts of outcomes will be incorrect. Any farmer who consistently uses an incorrect mental production function, for example, will end up with biased decisions. Deciding on appropriate fertilizer levels on a crop requires the knowledge of the response relationship. If this is several per cent above, or below, the real relationship, or the turning points are wrong, the fertilizer decisions will be consistently wrong (biased). In reality it is doubtful whether any farmer has the true production relationship stored away as this would require extensive trial work from his particular farm, but he should at least have a general idea using the results of years of observations on his and neighbouring farms, and from research stations that might be located in a similar situation.

How often have you made a guess? Farmers are inclined to make guesses when forecasting as this is easier than researching out what information might exist. If a farmer is inclined to do this on a regular basis, it is probable that the forecasts will be incorrect and, therefore, biased. The farmer's personality might be such that making guesses is part of what might be termed a *lazy bias*. This does not mean some farmer's guesses will always be incorrect. Indeed, some farmers may take in more information than appears to be the case and their 'guess' is actually based on considerable information and frequently relatively correct. However, the degree of guessing combined with a farmer's inappropriate intuitive observation powers gives rise to some farmers being consistently wrong. If the degree of incorrectness is sufficient, they will, of course, end up struggling financially.

The last well-documented forecasting biases are the *availability and recency effects*. There is a tendency to use data that are readily available simply because of its ease of access. Its accuracy is not checked and so its use may lead to errors. This could be called a part of the lazy bias. Similarly, farmers tend to remember and use data recently acquired in contrast to adding the recent information to the store of past information in order to create a longer-run set. Clearly this must be guarded against.

Overall, it is prudent for farmers to be constantly comparing the results of their forecasts with what occurs, and adjusting their procedures and mental assumptions and pictures in the light of the feedback. If this is successful, biased forecasting will be kept to a minimum.

People relationships

Managing a farm involves constantly dealing with people including the farm family members which may have both a managerial and physical input, either full- or part-time, through to lawyers and accountants who deal with the ownership, resource consent, employment aspects, through to taxation reports. In between are employees, neighbours and other local colleagues, commercial agents of various kinds, professional consultants, bank managers and contractors. All these people are the manager's interfaces to planning, getting the jobs done and meeting all the regulations. Good relationships are critical to success.

Stories of which farms to avoid when looking for work abound in some districts. A farm is often an intimate place of work in that the manager and employees are in constant contact and so if their personalities are opposed it becomes a difficult working environment. Similarly, successfully dealing with contractors who have many clients wanting their job completed immediately is crucial to success. While less frequent, ensuring the contacts with lawyers, accountants, consultants, bank people and the like are all appropriate and give rise to the outcomes desired is also important.

Unfortunately some people, even with the best of intentions, manage to upset others so that their dealings are less than successful in completing the job as required. Others are downright rude and cantankerous, characteristics that will not build a supportive and achieving team. Anyone who lacks all the people skills required can be said to have a bias in that they consistently end up with unsuccessful dealings with people relative to other farmers who have the knack of getting what they want achieved.

The whole field of human resources has spawned many books on how to succeed in this area, so the skill is in recognizing that there is a bias, and setting about solving the problem. Of course, this may be difficult where large personality changes are required. However, at least recognizing that there is a problem is a significant first step.

Also relevant is the manager's ability in assessing potential employees (*people bias*). It is relatively common to misjudge how a person will react in a range of farm situations. For example, someone who has experienced good luck might be deemed clever, a person with a dominant personality tends to be believed despite the facts, people try to act according to their assessment of what they perceive the manager would want, and similarly, potential employees tend to follow what is regarded as normal even though their beliefs might be different. When taking on staff, and indeed when judging people who interact with a manager, he must be aware of all these tendencies and make a judgement of the person's true personality.

Decision principles

Assuming that the information being used is correct (which it may not be), it is then necessary to use it logically to provide decisions and consequent action that achieves the objectives. Thus, any thought process that is illogical creates a bias. In general, without a good understanding of the production economics decision rules outlined in an earlier chapter, there is likely to be, on average, constant decision bias. Of course, one of the most difficult areas is the handling of risk and uncertainty. This requires the manager to be fully aware of the risk attitude which is part and parcel of the unit's objective function. This must then be built into making the decisions. Within this area, there are clearly many possibilities for bias including:

- incorrectly allowing for the risk attitude within a production economics context;
- using averages instead of marginal effects;
- ignoring the time value of money;
- using an inappropriate mix of inputs which do not produce least-cost production;
- ignoring the opportunity cost of unpaid inputs; and so on.

Overcoming these biases requires a good understanding of all the decision rules together with a careful analysis of the decisions made. Due to risk, uncertainty and time impacts, it is not a simple matter to come up with the right decision to compare with what was actually decided and therefore test for bias. Given the constantly changing technology and markets, years of comparable records will seldom exist.

Besides decision rule biases a number of other biases have been researched and recognized. These include:

- chunking errors;
- overconfidence;
- automation bias;
- 'lazy conclusion'(impatience);
- self-esteem errors;
- 'good follows the bad' syndrome; and
- 'market upturn hopes'.

A farm is a complicated production system which is difficult to wholly comprehend and mentally picture. Consequently it is common to break a problem into components called 'chunks'. The trouble is that many parts of a farm are interrelated thus making separating out independent components difficult. A farmer, for example, decides to compare all the crops he might grow and so calculates the per hectare profit from each by deducting the variable costs from the gross income. He might decide that he should, say, grow nothing but soybeans as they have the highest profit. The 'chunking' approach has been used in that each crop creates a chunk to be analysed for comparisons. In reality, it might be impossible to harvest all the beans when ready, as the available equipment and labour cannot cope when it all comes on stream, thus a proportion of the beans will be left too long and will deteriorate both in quantity and quality. If the whole farm aspects had been considered, it might be clear that the farmer should stagger planting dates, and/or plant a percentage in other crops, thus spreading the labour and machine load to produce a practical solution. Of course, allowing for the 'whole farm' would make the profit estimates more accurate. The lower gross returns would reduce the profit estimate for the whole farm being planted in beans. Perhaps different 'chunks' should have been analysed. A subset of crops might compete and form an independent unit that can be compared against other independent subsets and, therefore, can be considered separately in their own right.

Another common bias is 'overconfidence'. It is a human trait to imagine that we are more competent than actuality, though some people suffer from the opposite and always downplay themselves. This can be just as much a problem. However, if a farmer is consistently overconfident this may not impact on the choice of what, and how, to produce provided the options stay in their right order and it is similar for under-confidence. However, despite the correct ordering of alternatives, problems can arise as the overconfidence may lead to bad errors and disastrous outcomes. For example, confidently thinking that his skills will get him through a bad winter may mean a stock farmer will get very poor production and even animal deaths.

Related to overconfidence is 'automation bias'. There is evidence that humans tend to uncritically accept computer-generated solutions and information. This confidence can be totally misplaced in some computer systems.

A farmer's self-esteem can similarly lead to bias. A low esteem can lead, for example, to low market returns as the farmer accepts the low offers made for his products. It is easier to accept what is offered in contrast to checking the market and bargaining for an appropriate price. A low self-esteem might also lead to a very conservative approach to decisions and so opportunities are missed which, while having some risk, on average are likely to provide a higher return. Self-esteem can also influence people's relationships.

Impatience is another source of bias. Searching out all relevant information, and its careful consideration, is essential for good decisions. Yet how many people are somewhat 'slapdash' in doing this for they lack the patience to conscientiously and methodically follow through all aspects. Impatience will relate to a farmer's personality, so it is important for the farmer to recognize this and work hard at curbing the desire to proceed before the decision research has been completed. Impatience also impacts on day-to-day management. For example, a farmer may wish to bale the hay before it is ready despite a good weather forecast. Impatience might also be called 'lazy' decision making, though the root cause is probably different.

Overconfidence, an improper self-esteem and impatience can also all lead to the farmer misinterpreting the farm unit's objectives. This is particularly the case for self-esteem. As noted, getting the objective function right is crucial to success as without a correct yard stick very little will be right.

Finally, two other commonly observed biases need mentioning. Frequently you will hear a farmer comment that he will persist with a project although past conditions have produced bad outcomes, as this luck must soon turn to give good outcomes ('good follows the bad' syndrome). For some reason, if poor conditions have occurred farmers assume that the probability of better conditions must have increased. In the case of weather, for example, monthly events are likely to be independent of previous events. Bad does not create good, though in the longer-run, in the case of the weather, averages do tend to remain constant so eventually the 'bad' will be compensated.

Similarly, where a farmer has invested in a project, say a new intensive flower crop, where there has been a string of bad outcomes making the crop unprofitable relative to the alternatives, there is a tendency to incorrectly continue with the crop as it is expected there must be an upturn in the yields and markets ('market upturn hopes'). Unless there is very good evidence to support this, it may be better to cut the losses and move into another enterprise. This is all part of the syndrome 'after bad the good must come'. While this might be a good optimistic view to bring to life, it must be tempered with the real probabilities. This bias is part of the stubbornness of man. 'I thought of a good idea...it must come right so I can recoup my losses.'

Implementation

Many of the biases mentioned also apply to the implementation phase of any decision. Thus, overconfidence can lead to consistent bias, as can probability estimate failures, and an irrational faith in continuing with a project despite the poor outcomes. Probably two of the main biases are the:

- incorrect prioritization of the tasks involved in implementation; and
- the timeliness of actions.

Starting disease control well after the indications that it may be a problem will cause constant difficulties. Usually early recognition and handling of a problem gets the best response, so a farmer who constantly fails to notice the early signs will suffer relatively low yields. This is both an observation and timely action problem. Even if an emerging problem is noticed, if nothing is done to rectify it, the result will still be poor.

The other main bias in implementation is:

- a failure to constantly review plans.

In the ever changing and dynamic production world, conditions are usually different to what the norm might be, and similarly the prices and costs are seldom as anticipated. The logical response to these changed circumstances is to review the progress of the project to see if a change in the plans is appropriate.

In some cases, it will be optimal to continue with the original plans, but in many cases a change will be optimal. Thus, for example, it may now pay to purchase double the animal feed originally planned when the costs, and the feed produced on the farm, are both less than expected. The farmer must look at the situation from the point of view of the current condition of the project, and what the anticipated responses and prices are. It is from this constantly changing base, the new optimal decisions must be calculated. Sometimes these will be the same as before, and in other circumstances changes will be optimal.

It is the failure of some farmers to constantly update plans that gives rise to a bias. They must learn to accept that the decisions and plans made in the past are indeed history and should be forgotten because you cannot change the past. It is the future that must be focused on through a stock take of the current situation and the calculation of the next best move.

General

Finally, there is one other common bias that concludes the list of the best known. This is:

- an aversion to change.

This attitude is probably related to both a farmer's personality and past experiences involving failed initiatives. Some people just do not like changing their old habits to operate optimally for the current conditions. They have been introduced to a particular farming system, probably from a young age, and feel comfortable operating this particular system. They tend to continue with the system no matter what happens and hope that it will bring them through to at least a reasonable living. Of course, if the conditions change too much, they will end up loosing their farm.

There is something about such people that makes them feel very uncomfortable with change. Probably most humans have a component of change discomfort for it involves venturing into the unknown, and, therefore, creates challenges which some think they may not cope with. Clearly sticking to just the known will lead to constant bias in most modern farming environments. Farmers with this aversion to change clearly need support and encouragement from their families and friends as well as from the professionals who deal with them. Seeing others successfully making changes is also useful, and thus the value of neighbours getting together to compare systems.

Stress and Decisions

Introduction

As noted earlier, stress can give rise to biased decision making. A review of the literature makes it clear that stress is a very real phenomenon in agriculture, as is depression. In extreme cases, suicides result. Research in Britain concluded that 6% of the farmers had 'clinically relevant psychiatric morbidity' (Thomas *et al.*, 2003). This was similar to the general population, though a greater proportion tends towards suicide. Of course, while these extreme cases are serious and need timely intervention, the total cost of stress that is less extreme is significant for there is constant bias as a result.

Some farmers can be classed as 'relaxed' in that no matter what happens they simply get on with the job in an efficient and logical (unbiased) way. Others get quite stressed when situations are not as expected. Undoubtedly their personality is a major factor in how they cope with stress, as is their past environment. Many people will have come into contact with people who find problems in whatever happens due to the influence of people they experienced as youngsters. In many cases a really difficult situation does not in fact exist, it only appears this way to the stressed farmer. Thus, a combination of genes and environment experiences moulds how we will react in what some would call difficult circumstances.

Difficult circumstances can come about due to a range of factors, and some people react to some, but not others. The following is a list of possibilities:

- Weather impacts, particularly extreme and prolonged situations.
- Financial impacts such as the price dropping considerably for a product that has involved major investment, very high interest rates with high debt.
- Technological downturns such as animal health scares (foot-and-mouth disease).
- Overbearing regulations that require complicated paper work and reporting (health and safety requirements perhaps).
- Relationship problems, both on- and off-farm. It is difficult to escape situations on a family farm.
- Isolation and the working environment. Often farmers spend very long hours by themselves, particularly in bottleneck times such as harvesting, lambing/calving.
- Environmental pressures. Increasingly there is pressure from the general community over protecting the environment and complying with regulations in difficult situations.

It is a fact that dealing with the elements, world markets and people relationships including councils and governments in a small enterprise can be difficult, especially when your life's assets are tied up in the business, and it can be a fraught operation. Every farmer must be aware of potential stress situations and guard against the impacts of this stress (Fig.6.2).

Combating stress

To combat stress the person must be clear what causes the stress, and, of course, know what stress is. Stress will be created by any event, or anticipation of an event, that is seen as endangering the person's physical or psychological well-being. Thus, for example, when it is discovered that income will be much less than what was expected, and planned for, a stress reaction in the farm manager might well be created. In this case, it is probably quite clear what has

Fig. 6.2. Stress takes many forms. Overcoming such disasters takes support and strength of character.

caused the anxiety. However, what one person finds stressful might not impact on another in the same way. Where stress does occur, recognizing the cause is the first step in combating the debilitating impacts. For a full description of stress and its alleviation see, for example, Atkinson *et al.* (1999).

There are two general approaches to deal with stress:

- The first approach is to do something about the problem that is giving rise to the stress.
- The second approach is to do something about the mental reaction to the situation, and, therefore, reduce the impact of the 'stress' and the associated anxiety.

Combinations of the two approaches may also be relevant.

Modifying the situation can take several forms. Maybe the stress is caused by worries about meeting animals' feed requirements under a variable climate. If the potential feed problems can be overcome then the stress is removed. Possibly investing in irrigation where this is feasible may both remove the stress and increase profit. Or perhaps feed reserves might be built up to cater for the variable supplies. Or perhaps a positive bank balance kept and maintained to allow feed purchase when required. The advantage of cash reserves is that they can be used in any potentially stressful situations and, thus, remove many problems that might cause stress.

Other options for removing the problem include learning new skills (perhaps forward trading to cover exchange rate worries), using contracts to set the price received, and similarly contracts for physical tasks might help in 'bottleneck' situations. Then there is the opportunity to alter the mindset that creates the stress. Perhaps, for example, the farmer might review his objectives and consequently no longer worry about having, say, the highest crop yield in the district.

As noted, the second side of the coin is examining the farmer's mental reaction to the situations causing stress in an attempt to remove anxiety. The objective is to block, or change, the emotions that overwhelm the farmer and lead to irrational decision making.

One of the first things is for a farmer to keep records of the situations that cause stress. Some of these might seem trivial, but a full record helps to make it clear where the problems lie. Until this is sorted out, it is difficult to look for remedies. Perhaps, for example, a farmer is stressed over how much grain he might be wasting during harvesting. The solution might be to have an expert check how the machinery is set, which might both fix the problem, and set the farmer's mind at rest. Alternatively, maybe the farmer has to change his mental attitude to the problem in that, in the example given, the amount being lost is in fact normal and so the stress reaction was a needless self-creation.

One approach that many take is blocking out the problem. Sometimes a farmer will do this through alcohol, or even drugs, or simply through denial. None of these approaches is likely to help in the long-run, though blocking out a problem until the last minute before action must be taken can help with the idea that why worry about a situation that is not yet here, and currently requires no action. Sometimes blocking, or denial, subconsciously leads to unidentifiable concerns, even perhaps in a farmer's health status. Such a situation, if noted, probably needs professional help. On a more positive note, exercise can help modify the impacts of stress, particularly recreational exercise which provides further distraction from the problem.

A healthier approach is to share the problem with family, friends, and/ or neighbours, and also professionals such as a consultant. There is evidence that 'a problem shared is a problem halved' leading to less stress and rational solutions. The act of talking about a problem helps sort out a logical solution and once this is achieved the stress may well abate. Clearly this approach of finding a solution is by far the best strategy if it does reduce the stress for it may well also create a rational outcome. Some of the other strategies may not.

Of course, given any stressful situation it is also important to think it through for it might not have been avoidable, and there may be nothing that can be done. Something like an unforeseen market collapse clearly is not under the control of the farmer, and so it is important for the farmer to think this through and realize that it was beyond his control and so the stress felt, while understandable, is non-productive.

Another approach that may help in some circumstances is 'displacement'. This refers to altering, or replacing, a motive that has caused the stress. For example, a farmer may have an unrealistic expectation over the profit that is possible given his skill and resources. Perhaps the expectation can be lowered and in its place an expectation of greater leisure time created. This might involve an objective to lower his golf handicap! Effectively this strategy is one of changing the farmer's aspirations to a level that does not create overburdening stress. Another term sometimes used where a farmer's beliefs are different from actual behaviour is 'cognitive dissonance'. To remove the stress the belief needs adjusting to be more realistic.

On a more positive note, there is evidence that creating positive illusions can help stress coping (Taylor and Armor, 1996). The farmer needs to visualize the stress creating situation before it occurs, and then talk himself into being relaxed about the situation through bolstering his self-image and perceptions of control. Effectively the farmer is simulating the potential situation and working out how to handle it under the assumption that he has all the skills necessary and it will all be a smooth operation without any stressful outcomes. Certainly in the sporting arena there is significant evidence that visualization of success does in fact assist success occurring.

Another positive approach is for the farmer to learn relaxation techniques, and to ensure he has plenty of aerobic exercise, both of which can reduce the impact of potential stress. A relaxed person is in a state opposite to that of a stressed person. There are many available courses and books on relaxation techniques (other than counting sheep, or some other object, when trying to get to sleep!). Most techniques involve working on relaxing the muscles, from toe to head in that order, ending up with a consequent mental relaxation.

Finally, the farmer has the choice to avoid situations that cause stress. In effect, this might be achieved through some of the strategies already mentioned. For example, use of forward contracts might well remove stress caused by variable income, reducing the number of animals farmed reduces the feed supply stress, and so on. There are myriads of strategies that can be used to dampen the stressful situations, each one of which should be examined in the search for a better system.

It should be noted that dangerous stress situations may need professional psychiatric help. Anyone involved with farmers in an advisory/consultative capacity must be very careful to seek such assistance whenever there is real concern over the health of a farmer. The farmer's family will more than likely be similarly seeking assistance, though sometimes it is easier for an outsider to organize this. If the farmer is in denial, the situation is that much more difficult.

It should also be remembered that within the farming population there is a continuum of stress reactions and feelings right from the extreme level, referred to above, to the very mild which has no harmful bias impacts.

In summary, the main techniques discussed here for reducing stress involve:

- keeping records to first identify the causes of the stress;
- altering the system giving rise to the stress;
- learning new skills allowing the use of stress-reducing techniques or approaches;
- use of risk-reducing techniques such as forward contracts, diversification and the like;
- changing the objectives/attitudes (called displacement);
- sharing the problem with friends/relatives/professionals;
- positive illusions;
- relaxation techniques; and
- avoidance of the stress creating situations.

The Case Farmers' Thoughts on Bias and Stress

Margrave's case

Margrave certainly experienced some stressful times. One example involved a rival for a position where the competitor was a local person. The family received threatening phone calls making life very difficult for a time. But in the end, the people skills Margrave exhibited meant he eventually developed a good rapport with the person and often chatted to him about farming issues. Perhaps Margrave had many intuitive skills which automatically came into play. He recounts the story of buying a holiday house, or rather the attempted purchase of the house. It was in an idyllic spot near the sea with a facility for Margrave to store one of his few off-farm passions, a trailer yacht. The night before the auction Margrave had trouble sleeping and woke several times, as he commented: 'one thing I never did was have trouble sleeping'. In the end, he decided against attending the auction and later discovered that the historic house was on a spot that had once seen a massacre of local Maoris who were ransacked by a visiting chief and his warriors. Can this intuition be explained?

Margrave believes he has a bias over being too kind to people and does not dismiss or reprimand staff when perhaps he should, but despite this bias he certainly seemed to be successful with labour relationships. Margrave also believes he is perhaps insufficiently anxious on occasions when he should be worrying and sorting out an issue. However, he always does an enormous amount of homework and so there is seldom a need to worry for he has confidence in his conclusions. In the end, he walks away from decisions and situations he is not comfortable with. Is this a bias?

In his desire to have high production, Margrave believes he has too many animals and is overstocked at times relative to the production. He noted: 'I have rosy tinted spectacles' and 'should learn to "back off"'. Margrave is the total optimist and thinks that the next spring will always be one of the best (particularly with respect to animal feed. See Fig. 6.3). This 'does get me into trouble'. To counteract this optimism he keeps a large photo of his farm in the worst drought pinned to his office wall.

Profit was always at the back of Margrave's mind and forecasted outcomes influenced decisions. However, on a daily basis 'financial aspects don't influence daily management and I probably don't do enough cash flow forecasts...however, the situation has never got bad enough to create depression'. Margrave always discussed most situations with his wife and consultant and so this sharing may well have helped prevent financial stress. And to further help his outlook, both mental and physical, Margrave always moves his stock around so the areas near the house always appeared green and 'this helped morale enormously'.

The comment about cash flow budgets is also reflected in his attitude to records. He notes: 'I enjoy the technical aspects of farming, but I certainly don't like keeping records and monitoring. Records...what for?' However, Margrave always weighs animals to assess their condition, and carefully notes feed reserves with the outcomes being stored in his mind for 'instant access'. This reflects his excellent memory and his confidence in his ability to assess all relevant facts without the stress of keeping endless records.

Fig. 6.3. Feed levels are always a major source of stress, particularly on hill farms like Margrave's where irrigation is very difficult.

Now Hank's views

Hank also recognizes that he does have some biases, and certainly experiences stress at times. It would be surprising if good farmers never experienced stress for part of being excellent is being human with the associated emotions and drives.

Hank comments: 'I'm just no good at formally monitoring what happens round the farms...I find it boring, and what's more my monitoring skills are terrible. For example, no matter how hard I try I'm useless at judging the dry matter level of pastures, and equally no good at accurately judging the live weight of cows. What's more, I'm not only well out, but my estimates are inconsistent.' Thus, Hank relies on staff to carry out these monitoring tasks. Hank freely comments that he finds the planning and execution side of management challenging, fascinating and enjoyable. This is where he spends heaps of time. What Hank recognizes is his areas of bias and weakness, and so adjusts accordingly. This is clearly very healthy.

Hank also recognizes that he has 'rosy tinted spectacles' when it comes to estimating feed costs with understatement being common. Hank comments: 'I always underestimate input prices in general.' However, he updates his cash flows every 2 weeks so that the underestimates seldom have time to lead to major problems. Of course, these are self-recognized biases which are compensated for. The real problem is the existence of other biases which are not corrected. This is where a third party can help, or perhaps family members, or even trusted employees.

Perhaps rationally, Hank also believes he is biased against intensive high-input-level systems that require housing the cows for long periods. He maintains there is insufficient extra profit for the risks involved. Small decreases in product prices leaves you teetering on losses, and with such intensity, a small error or disease problem causes havoc to the profit. Perhaps he is right.

This attitude, which Hank calls a bias, is perhaps related to his attitude to risk and stress. He just does not want to live on such a knife edge. Hank recounts how he does get stressed at times of the 'nasty hot dry weather'. He can remember one season when the irrigation well was only producing half as much water as normal and so he was up day and night shifting irrigation systems to get the most out of every litre he could pump. It became a vicious circle with middle of the night panic attacks. The situation had got the better of his rationality and emotions. Help from others finally led to thinking through the problem and the return of rationality. 'It was pretty scary.' But it was also a learning curve about putting circumstances into perspective and acting appropriately. Hank would not be the first farmer to face such problems, particularly exacerbated by high levels of debt.

Concluding Comments and Bias Reduction

Most farmers exhibit both some form of biased decision making and stress. The latter, however, may not impact on efficient decision making, but where it does, remedial action is desirable, just as it is for non-stress-related biases. The main difficulty is recognizing the bias that exists; for once recognized and annotated, correcting the problem is likely at least to a certain extent. Correcting overbearing stress, however, may be more difficult.

There are three approaches to recognizing bias. The farmer himself needs to be constantly aware of the possibility of bias so self-examination is the first approach. The second is to employ other people to consider the possible biases and provide suggestions as to where improvements might be possible. For the farmer to employ other people requires a certain amount of humbleness and a willingness to accept an expression of poor management as well as a willingness to change. Third, it is conceivable that formal tests might indicate areas of bias. Currently such tests are not generally available, but, for example, it should be possible to present well-described and well-illustrated decision problems for which the farmer is asked to provide the answer. If the simulated decision situations are reasonably realistic, it should be clear if bias exists. Of course some farmers will not manage written tests as it can be hard to relate paper situations to reality. Currently, however, bias recognition must be human-based.

For self-assessment, it is useful for the farmer to have benchmark data on the performance that might be expected. This should ideally be from other farms in the same environment and start with profit information. Detailed benchmark data can go down to crop yields for different input levels. Such data need to be carefully collated and allow for seasonal weather variations through having data for as many years as possible. One year's data are unlikely to be useful for random impacts can influence the comparisons. It must also be remembered that the data might not reflect the ideal for while some farmers might perform better in some areas, they may well be worse than the case farmer in others. Thus, while a comparison is essential, it must be used intelligently.

Any comparison process must take cognizance of the farm's objectives, for profit and cost might not be the main priority, though physical input/output ratios are likely to be universal in that few would not want maximum output for given inputs. Again, however, production economics principles must be allowed for in that, for example, maximum output per technical unit (e.g. hectare) is unlikely to be optimal.

Where outside people, or even family and neighbours, are asked to help, they will similarly have benchmarks in mind to allow some sort of comparison and conclusion. Some of these benchmarks will be related not only to their experience of other farms, but also of other managers. This is the additional judgement other people can bring to the assessment relative to a self-assessment. While such judgements will be subjective, an experienced person will certainly have ideas of where the bias occurs. One obvious area is how the manager deals with other people. While in the end there must be opinions involved, a good manager will welcome constructive ideas as food for thought and possible action.

Little can be said about the formal use of tests as they are yet to be developed. But examining how professional consultants assess a farmer would lead to ideas of what might be included in such tests. In that personality and management style are more than likely related to biases, these tests will be of assistance in assessing a farmer. For example, a judgement on a farmer's anxiety personality trait can be particularly helpful in assessing stress levels. Other traits will similarly influence outcomes, another example being the farmer's attitude to new ideas (openness). As bias is related to personality, style, experience and training, the last chapter (Chapter 5) contains a more detailed discussion about how these factors might be altered to give greater management skill. After all, this is the main objective in considering a farmer and his attributes.

Suffice to comment at this stage that probably one of the most important skills in noting, and working on improving, biases is the ability to be self-critical. By taking a critical approach is not meant as being negative about everything, but being objectively analytical in looking for biases and working on their correction. Furthermore, of course, a critical approach in this positive sense is the most important attribute for all areas of farming. When, for example, a commercial agent tells the farmer about the large number of positive attributes the product he is selling has, each should be carefully examined to assess whether the benefit would be achieved on his farm. Effectively, everything should be taken with a grain of salt until clearly proven.

As part of this critical approach, all decisions should be double-checked, and where possible discussed with others for this process clearly outlines what is proposed, and why, and, therefore, leads to both a self-review as well as others' input.

In the end, change and improvement must come from within.

7 More on Objectives: Family Influences, Origins and Modification

Introduction

A farmer's objectives strongly impact on the decisions made. This is one of the reasons why the outcomes from every farm tend to be different as each farmer's objectives will be unique. Also important to decisions is the farm family including a spouse. Thus, objectives and families are further considered in this chapter.

Part of comprehending a manager is the understanding of his or her objectives, and the origins of these objectives. So, one of the first steps in helping a farmer is determining whether his objectives are correctly stated. Progress cannot be measured without these yardsticks. But, while determining the objectives (perhaps using the questionnaire listed earlier, or through careful observation) is important, of even greater value is the understanding why the farmer holds the particular set. Possibly the farmer has concluded incorrectly and so discussion and assessment may lead to modifications. To this end one of the sections in this chapter contains a discussion on the objectives and the influence of the family. Similarly, as a farmer's locus of control (LOC) may be important in constricting progress, factors which give rise to a particular attitude are considered with a view to understanding his LOC, and what might be done about improving the situation.

Everywhere in the world most farms tend to be both owned and managed by a farm family. Thus, an important influence on the farm operation is this family, both spouse and children, and in some cases, even the earlier generation. Accordingly a discussion on their influence is included in this chapter. The family is part and parcel of the human factor.

When talking about objectives, the question of entrepreneurship should also be considered. In an earlier chapter, motivation was referred to with the comment that it is encompassed by managerial style rather than being an independent factor. The same might be said of entrepreneurship, but as it is a much discussed subject it will be further considered.

A farmer who is at the forefront of innovations and always seeking new products, methods and 'off farm' ventures that might perhaps build on his farming success is often referred to as an entrepreneur. Such farmers hope to better achieve their objectives by being adventurous and creative in their business operations. A stock farmer might, for example, buy a chain of specialist butcher shops to market his, and his colleagues, produce in the hope of making better margins, and increase his wealth and power. What are the features of such people, and can others acquire these features where they aspire to expand?

The Family Influence

Family relationships are critical to the success of a farm. Farm business is largely unique in the business world in that the family is intimately involved through both living on the farm, and probably being part owners. Very few urban businesses face this situation, which is both an advantage, and in a few cases, a hindrance. Farmers and their families often live in quite isolated situations relative to urban situations, and for many days a farmer and his spouse will have no contact with other people, and similarly any family member working on the farm. This is an important reason why the family must be in harmony, and, if not, achievement of their objectives will be made difficult. And while corporate ownership of the farms is increasing in many parts of the developed world, it is still in the minority as many so-called company farms are in fact family-owned and operated. For example, in Colorado, around 90% of the farms are family-owned and operated. And in New Zealand the ownership is held by:

- The farmer in 30.5% of the cases,
- Family partnerships 57.5% of all cases,
- and family corporations, trusts and public company farms for the remaining 12%.

Where there are truly corporate situations with the owners totally divorced from the management, the family influence is largely non-existent, with the manager simply carrying out the board's instructions on the objectives and the goals. However, the majority of farms are not in this position, and it is doubtful whether this will ever be the case, as most farm families are prepared to accept a lower return than that required by corporate investors. Furthermore, farming is a 24 h, 7-days-a-week operation and there are few employees happy to work such long hours without an intimate involvement in sharing the benefits and costs. For this reason, sharing some farming arrangements can be successful as the manager has a direct interest in, say, shifting the stock in the middle of the night when a flood is imminent. To achieve success, living on the farm is often a prerequisite. Truly, the human factor is crucial in the life of a farm.

With family farms there is competition for the:

- resources;
- cash; and
- time.

How much of the available time is devoted to production, and how much to the family and leisure? When the products are sold, what is the allocation of the proceeds to costs, farm improvements or maintenance, and what percentage ends up on consumption and the family home? Clearly there can be considerable conflict between a spouse and the farmer, and the children, not to mention other family members who might be involved such as a father, or working children. Family relationships and procedures are clearly important to ensure harmony and good decision making for both farm and family (Fig. 7.1).

In by far the majority of cases, a spouse will be a wife, and at some stage in the life of a farm, a mother. While there is a trend for females to be farm managers and owners in their own right, the percentage is still very small. But that does not mean the spouse is not involved in both farm work and decision making. This participation also means a spouse is increasingly involved in setting objectives, though often this is not directly expressed in setting out a list, but rather implied in the decisions made, such as devoting money to upgrading the farm house or on education or perhaps leisure activities. The survey data suggest that spouse involvement in a range of decisions covers the full spectrum from complete involvement through to none at all.

While many families work harmoniously over objectives, decisions and relationships, it is certainly not always the case. Harmonious relationships depend a lot on:

- the personality of family members;
- the interests of the members of the family group; and
- how well they listen and consider each other.

Fig. 7.1. Competition for resources between the farm and the household can sometimes be intense and problematic.

The success of the farm operation does influence the relationship and if all outcomes are considered at least satisfactory, there are less likely to be tension spots. But where outcomes are well away from what might be expected, this can erupt in poor relationships and a dysfunctional situation that only outside help, or valuable leadership from one of the family, can solve and lead to better outcomes. Thus, an observer must be careful to assess not only the farmer, but the whole family unit and how it cooperates and works positively.

Harmonious families will have:

- good communication skills; and
- open relationships.

They will be constantly talking to each other about their wants and needs. Part of this communication will involve all members being encouraged to express their feelings in a totally accepting way. Each needs to know, and consider, the others':

- needs;
- desires;
- attitude to power; and
- where they see the future evolving.

Similarly, their thoughts on the best course of action for each problem need to be fully canvassed and considered to finally arrive at a consensus.

Of course, not all families will need to work this way, as in any population, there will be a full range of family situations and power distribution. Depending on the members' backgrounds and beliefs (more on this later), some families will be happy for the farmer to be the sole decision maker in an autocratic operation. Increasingly, however, such situations are decreasing with the increased liberation of spouses and improving education opportunities. The family itself must work out, perhaps with help, the most suitable arrangement.

Some of the potential stress points include:

- a farmer not considering the impact of decisions on each family member;
- a period of poor prices and weather outcomes, all giving rise to financial stress with few ways out of the situation; and
- the family members having very different objectives, goals and needs, and these may all have diverged since the farm was set up.

The intergenerational transfer of resources is always a situation fraught with potential problems over fairness between siblings, though careful thought, discussion and professional help can lead to family harmony. If parents remain on a farm, it is hard to develop harmonious decision making with the son or daughter taking over. Different generations have different upbringings, agendas and beliefs that seldom match. This is an age-old problem existing in all parts of modern society where change is considerable. Integrating with daughters/sons-in-law may not be easy and requires a benign approach by the older generation moving to a different phase of their lives.

The list of potential family stress points is almost endless. But many families do work well and meet their objectives with satisfaction and harmony. Success must depend on an awareness of each person's:

• strengths;
• skills; and
• weaknesses.

This applies to all individuals in a family, so that they can pass on a problem to the person best able to handle it. Goal setting and time management are all important. Tasks not completed in time lead to enormous stress, and in biological production, time waits for no one with, for example, the weeds seeding, the rain moving on, the market peak passing, no matter what the farmer does. Thus, priority lists and realistic time estimations are important aspects to decision making.

Many believe that democratic decision making is important, if not efficient in the time taken. Involvement leads to commitment. And particularly in the most difficult times, it is important for all members of a family to have hobbies and recreational activities, including being part of the local community, not only for enjoyment, but also to help keep life in perspective. The old saying 'all work and no play makes Jack a dull boy' still has credence in more ways than one.

Despite the best of intentions, dysfunctional families do exist. The observer and the consultant must be able to recognize this and carefully help resolve the situation. Of course, in some cases, the situation is irretrievable, in which case the best course might be a major reorganization. A dysfunctional family can often lead to poor decision making and the non-achievement of goals and objectives for all members of the family. In these cases, it may be prudent to suggest professional help be called in, though this is difficult to achieve if the family does not recognize that they have a problem. Many organizations, and similarly professional counsellors, are available to help. Much literature is also available (search the web, talk to community leaders, etc.).

The family problem may relate to the married couple, or involve the children as well, and even the previous generation. Each situation will require different expertise to help. A counsellor should be carefully selected to suit the personalities and beliefs of the people in conflict so that the discussions and contacts are important before selection. Most counselling involves a series of steps, which are followed through with the help of the professional. Seldom will a professional simply provide advice and instructions as solutions must come from the participants themselves. The professional will guide such resolutions and provide support. Usually an imposed solution simply does not work.

The steps often involve the following:

1. The individuals are asked to list out the good points in each other.
2. They are written down before telling each other the list in a very positive way.

Then, hopefully, the following is possible:

3. Each of them lists out the actions which she or he believes harms the relationship.

These steps require the participants to really listen to each other and absorb the feelings. This is where the counsellor can play an important role, because very often people do not want to hear about the 'not so good things' and brush them off. A real understanding of the other members' concerns and feelings is the first step to resolving issues. An acceptance of the 'no blame' approach is critical to the next step.

This involves taking responsibility for the concerns, and working out solutions that are mutually acceptable. Resolution is under way. Even without counselling, these open communication procedures can be practised. In either case, key words are:

1. 'Listen'.
2. 'Learn'.
3. 'Set aside time' to do both.
4. At all times practise remaining 'cool, calm and collected'.

This is where a third party can help make real progress.

Of course, counsellors will have their own variations on this basic process and resolution. And each family will require a slightly different approach, and the process will be dynamic taking its own individual course. Some families may prefer to tackle the situation by themselves, and certainly there are many good resources available. The local general medical practitioner may well have suggestions.

The emphasis in this discussion has been on tackling dysfunctional families. But families fall on a continuum of dysfunction with few being totally harmonious from a management point of view. Thus, the seemingly well-adjusted family may well obtain benefit from going through the procedure suggested.

Bruce the consultant has firm views that a consultant can be an important third person in helping families better integrate themselves into the totality of the farm and the family. He has observed that dysfunctional families have a major impact on the farm's success. And particular life stage events can be concluded successfully with a consultant's help. Two stages of notable importance are when a family member, most likely a son, returns to live and work on a farm. Usually significant changes are necessary to provide the income to support an additional person, and often a shift of power must eventually be organized. The other important stage is when the parents are ready to retire. But planning for this must start very much earlier and can benefit from the consultants expertise, especially if she or he knows the family and the personality of its members.

Margrave's family was certainly not dysfunctional. Both Margrave and his wife (perhaps we can call her Margaret) were born in a middle-sized rural town on an idyllic coast. In Margrave's case, the family farm was not far away and so it was convenient to take the trip into town. And later on Margrave's family, now including two boys, bought a house in town where they all lived so that the children would have easy access to education. Margrave then travelled daily to the farm where most of the staff lived ensuring proper cover. Besides, each staff member had his area of responsibility. Initially Margaret helped a lot on the farm, and was involved in decision making, but once the move to town occurred the involvement was much less. Besides, Margaret, who is a university graduate, soon took up further distance education and had many challenges to occupy her mind. The boys in their earlier life made full use of the farm as an exciting playground, and in later years worked on the farm carrying out jobs like cultivating large areas for growing winter feed. They were hungry for the good pocket money on offer and were happy to work long hours on this boring job.

It is clear the family influenced decisions regarding their living situation with education being paramount, but due to the later separation of their lives from the physical location of the farm they were not heavily involved in the immediate decisions. And with Margaret developing her own professional expertise, her involvement was similarly somewhat divorced from day-to-day decisions. And now the boys have largely left home having been through various training programmes leading to one being a project manager and the other involved in survey work. Neither has a strong desire to become involved in agriculture and will probably make their way in urban life. But who can see the future?

In Hank's case, the family had a significant influence in their progression through agricultural life stages. Hanna was born in a rural area, as was Hank, so rural life was part of the family tradition. Further, both had strong agricultural and rural ambitions giving rise to clear objectives about creating a strong farm and business, and in that they started with very little, they have been most successful. Clearly, shared views on their direction in life make the sacrifices and commitments so much easier.

As happens in families, the children were an important influence on resource allocation both from a daily living point of view, and from an educational view. Both children were encouraged to follow through with tertiary education, and achieved these goals successfully (one graduated with a degree in Horticultural Commerce, and the other has completed diplomas in hospitality management). Now, while their daughter is currently at home, temporarily before departure to another country, shortly it will be just Hanna and Hank at home. This completes the standard life cycle requiring Hanna to alter her objectives. This is not always easy as for so long the family objectives were very clear with farm and family being dominant in their struggle to develop resources and net worth. Hopefully, soon the re-evaluation will be complete setting forth new directions for the next stage in their lives.

The Influence of a Spouse

By far the majority of the farms will involve a partnered couple, and by far the majority of the spouses will be female. Thus, anyone involved with working with farmers to improve their management, or even to just observe what is happening on a farm, must consider the spouse's personality, intelligence, objectives, as well as family relationships.

Feldman and Walsh (1995, p. 39) noted: 'This division of power, control and skill shapes the decision process.' Clearly a spouse can be a powerful addition to the management team, both in a decision sense and in a farm work sense, for an extra hand can be crucial in some tasks. Just how much a wife/

partner becomes involved in farm operations varies enormously across individual situations and cultures. In Queensland, for example, it was found (Rickson and Daniels, 1999) that:

- 51% of husbands believed they made crop decisions by themselves; and
- a further 30% believed that they tended to have more influence on crop decisions relative to their wife.

In contrast, for household equipment decisions:

- 49% of wives believed that the decisions were shared equally;
- a further 33% believed that they had a greater influence than their husband; and
- the remainder of the wives believed they made the decision on their own.

On the other hand if you ask the husband, they believed that 25% of the wives made the decision on their own.

The figures for other decision types range round these quoted figures and so it is clear that, overall, the farmers do tend to have a greater influence than their wives in decision making, but this does vary from case to case as shown, for example, by 19% of the husbands believing crop decisions are shared equally.

Each family works out what they are comfortable with regarding the lead person in each decision. Of course, where the relative responsibility is in dispute, problems may arise. And these decision relationships are not static in that as a family evolves the decision-making responsibilities change. In times of financial stress it is often observed that a wife's interests suffer in favour of maintaining the farm, and similarly the decision-making involvement suffers.

Another study (Sawer, 1974) found that wives who spent time finding information and data were more heavily involved in farm decision making, as were spouses who spent time physically working on the farm. In this situation, the wife probably made an important contribution to the decision involved. Sawer also found that the more children the couple had, the female had less involvement in decisions, and there were similar negative relationships in decision involvement with respect to socio-economic status (income, farm size, education, age, social participation).

A farmer's adoption of innovations was also related to the spouse's decision involvement, so perhaps this involvement helped share the risk and the responsibility leading to the farmer being prepared to take on the unknown at an early stage. It is also interesting to note that the wives in Sawer's study had, on average, 12 years of education, whereas the husbands had 8 years suggesting that many wives contributed through their better education.

In a study of 880 Kentucky farm women (Bokemeier and Garkovich, 1987), the researchers found that women who had been raised on a farm were more likely to be involved in farm decision making and tasks, and tended to have a strong self-identity regarding the farm situation. This is a similar situation to managerial skill where it was shown that early farm life was important in developing good farm management skills. It is clear that all the work in the general population on the influence of early life on subsequent success appears to be similarly played out in agriculture.

It also appears that the household and the farm are often in conflict over resources, bargaining sometimes occurs over finding a compromise…"if I can do 'this', then you can have 'that'." This is particularly evident when the debt loads are high which in turn leads to stress. No one likes constantly increasing the debt load when conditions are difficult.

All the information presented makes it clear that the spouse is a very important member of the decision team. In helping improve decision skill, just as much attention must be given to the spouse, and similarly when introducing technological improvement it should be ensured that the information is available to the spouse who should also be involved in any discussions, and, where possible, demonstrations.

A wife's self-identity is usually enhanced by involvement in decision making and farm operations. There can be nothing worse than living in a relatively isolated situation and being regarded simply as a household keeper, though usually a wife is heavily involved in leisure decisions and activities thus leading to some satisfaction (Fig 7.2 suggests craft can be a source of fun). Family decisions usually also provide satisfaction and identity.

Overall, for success and harmony, it is important to evolve good decision systems with a wife contributing towards many aspects through discussion and resolution. Furthermore, a spouse can be actively involved in the decision steps

Fig. 7.2. Spouse involvement can take many forms. In this case another source of income as well as community involvement.

from finding out information, sorting out goals, estimating costs and returns and resource requirements, and finally helping make the decision to proceed. It is interesting to note that on average women probably work longer hours than their male partners. One extensive US study (Kim and Zepeda, 2004) found that women work on average 13.8h per day over 7 days, whereas their partners worked 13.5h. This includes household tasks.

Working in a harmonious and beneficial way as a wife/husband team assumes that the marriage is successful. This is not always the case as shown by contemporary statistics (in the USA something like 50% of first marriages end in divorce, 67% of second marriages and 74% of third marriages). The rate in different countries varies markedly, as do the number of official marriages per 1000 population. In New Zealand, the rate is as low as 14.6% and lowering with many de facto relationships. While it is not always easy to predict which couples will evolve a successful marriage, research does give some pointers. Understanding the personality traits which lead to success is important as it gives an idea of which couples are unlikely to be able to improve markedly, and in the other cases, which characteristics need targeting. Of course, helping dysfunctional couples is a specialist task, so any attempts in this area must be treated with caution.

One study (Bentler and Newcomb, 1978) found that personality was a successful predictor of marriage and, therefore, whether a couple might be able to sort out their difficulties. Similarly, Kelly and Conley (1987) found that the neuroticism (anxiety) of both the husband and spouse were important predictors as well as the impulse control of the husband. This study followed 278 married couples over 45 years. Twenty-two engaged couples did not make it to marriage of the 300 starting in the study.

Many studies have found similar results providing confirmation that the personality of the couples is important to a good relationship and, therefore, an increased likelihood that they will make good farm decisions. For example, Caughlin et al. (2000) used a 13 year longitudinal study in which they measured aspects of the personality of married couples to assess correlations with marital satisfaction. The two most important traits they used were 'trait anxiety' (apprehensive, tense, emotionally labile, suspicious, shy and undisciplined) and 'negativity' (e.g. spouse showed anger or impatience by yelling, snapping or raising voice).

A negative relationship between trait anxiety and both marital satisfaction and communication was found.

A positive relationship existed between an:

- individual's trait anxiety and negativity;
- between the wife's trait anxiety and the husband's negativity;
- between both the husband's and wife's marital satisfaction and the spouse's negativity; and
- the husband's negativity was related to the wife's marital satisfaction.

It was also clear that spouses were directly affected by their partner's emotional makeup. The implication from all these studies is that working on improving aspects of couples' personalities is likely to improve decision efficiency as well as marital satisfaction. As noted in the following chapter on skill improvement, there is good evidence to suggest that personality (management style) can be altered given good support, a strong desire to change and skilled counselling.

But for Margaret and Margrave, it appears that marital disharmony was not an issue in their lives. Margrave could devote all his energies to making good decisions knowing home support was always available. Margaret has been, and is, involved in a range of jobs, and besides a university bachelor's degree, also has training in school teaching, business administration and also achieved a university master's degree. Some of this training was through distance education reflecting a determination to succeed. Employment has involved professional jobs in both rural and urban occupations, and still does. All this experience and training has enabled Margaret to provide excellent support in all decision and analytical problems related to the farms. This is not uncommon in rural western life with highly intelligent spouses looking for outlets for their talents and for satisfaction in making important contributions to the community. The team that many rural couples develop is powerful once the children are less demanding. And prior to this stage, the assistance and example provided to the children is similarly powerful and motivating. And of course there is the extra pair of hands provided at crucial stages. The life cycle described here is an example of what the literature reviewed reveals.

And Hanna has contributed to the farm development in Hanna and Hank's case. Hanna graduated with an Agricultural Science degree and was, and is, highly qualified to be part of the decision team. Now that their children are more or less independent, Hanna has become heavily involved in the recording and accounting side of the business and constantly produces critical financial reports. Prior to some health difficulties, Hanna also took charge of raising the calves, a task which she loved and gained considerable satisfaction from knowing that the results led into valuable herd assets. Hanna has also contributed heavily to the rural community and over the last few years has been the president of the local golf club ensuring its continued place in the sporting and social life of the district. Others might involve themselves in church activities (Fig. 7.3). Rural people are very aware that they depend on each other for so many parts of their lives. That is what makes a country.

Bruce the consultant is also under no illusions about the importance of a spouse in the overall scheme of things. Bruce comments: 'A consultant must work with the spouse in most cases for she has a major influence in what happens'. Sometimes the spouse is from a long-standing farming family, who may well have provided resources to help the 'young couple', and consequently rightly believes she is an important part of the system who has considerable knowledge of the farm. The consultant, logically, must deal with both members of the team.

The Origins of Objectives

As often stressed, a lynch component of farmers and their families are their objectives which drive and underpin all decisions. It is crucial to understand

Fig. 7.3. Contribution to the local community is often an important part of rural life, and can take many forms.

both their nature and origins when working with farmers for otherwise any discussion on possible changes is likely to be misdirected. Given the stimuli being received by the farmer, such as next season's price predictions, the objectives should dictate the management response. To ensure an appropriate response the farmer must clearly understand the objectives both he and the family hold. As Hobbs *et al.* (1964, p. 1) noted 'objectives set the goals, and strategy sets the path to the goal'.

As noted, there are many sets of questionnaires in the literature for assessing objectives including the one listed in Chapter 3 (which the reader may wish to review). In this literature, different terminology is often used. Examples include the terms 'values' and 'goals' as well as 'objectives'. However, when you think about it all three words can logically be used.

Every farmer, and the family members, will have a set of basic values, such as a belief in being independent and secure, which they wish to adhere to. These give rise to their objectives. For example, wanting to be independent and secure probably leads to an objective of reducing debt. In turn, the goal may be to reduce the mortgage by 15% over the next 3 years. Thus, values lead to objectives which, in an operational sense, give rise to specific goals and targets which lead to actual plans and forecast budgets.

You probably have a wide range of values which guide how you react to different situations. Similarly farmers and family members will have developed value sets resulting from their background and experiences. The set will be heavily influenced by the early family environment, and similarly the community in which they live, as well as the general culture encompassing both the family and community.

Examples of basic values include:

- the degree of security sought;
- the drive to conform;
- the level of power sought;
- the benevolence level;
- friendship;
- loyalty;
- forgiveness;
- helping others;
- politeness;
- attitude to equality;
- the respect required (community status);
- honesty and openness;
- anti-pollution by both natural substances and man-made substances;
- sustainable production and organic production values;
- leaving the farm in a better state;
- independence;
- importance of innovation (creativity);
- valuing education for its own sake;
- balancing work, leisure and comfortable living conditions;
- working for an adequate living and conditions compared with maximizing;
- not working on Sundays relative to religious beliefs;
- ensuring full use of abilities and resources as a national responsibility.

And so the list can continue.

Not all values will necessarily impact on farm objectives, but in that a family farm intrudes into most parts of a farmer's life and times, and most will impinge on operations at some time or level. Values such as honesty are all pervasive, and even values such as emphasizing education will have some influence on the objectives as it may, for example, be necessary to curtail farm comforts to provide education funds.

In an important early review, Gasson (1973) summarized the wide range of possible values into four main categories. She called them as follows:

1. Instrumental, e.g. farming as a means to an end – income, security, congenial working conditions.....
2. Social, e.g. being part of the farming community, good relationships with employees, family relationships.....
3. Expressive, e.g. self-respect, creativity, overcoming challenges.....
4. Intrinsic, e.g. value of independence, enjoying outside work, family involvement.....

Gasson reported on a range of farm surveys and concluded that intrinsic values were particularly important to most farmers, but that the farmer's situation also correlated with the importance of the other main categories. For example, low-status farmers particularly valued independence, and the social values were important to middle-status farmers. In contrast, the higher-status large farms had managers who valued instrumental values. They had moved into being serious businessmen/women striving for maximum profit.

However, it must be stressed that categories can be dangerous and each farmer has a unique background and situation that gives rise to his particular set of values, objectives and goals. And it is also worth noting that the research reported concentrates on the farmer's values in contrast to the total family values, though in some cases the farmer's values will reflect the totality of the farmer and the family. In modern times any research should consider both the farmer and his spouse, and possibly other family members to. As the years have passed the farm spouse has become an increasingly important part of the management team.

Values, objectives and goals are seldom static, so anyone considering farms, farmers and farm families must constantly review conclusions and associated family support. For many years farm management studies assumed the objective was to maximize profit, and farmers not reaching this nirvana would be regarded as irrational (some, for example, might find nirvana in fishing... Fig. 7.4). In reality, it was the commentator who was irrational for few farmers have a simple objective of maximizing profit. Indeed, many might say today that most farmers are rational; it is only the inability to define their goals that makes them appear irrational. This is probably not true in many cases for efficiency studies that relate one farmer to the others show a significant proportion could improve output with the same resources, or maintain output with fewer resources. Again, some of these conclusions probably stem from an incorrect assumption about what criteria to use, but certainly not all farmers

Fig. 7.4. A farmer with this resource bordering his land is truly lucky, and who would blame him if it influenced his objectives!

believe they are competent and rational as evidenced by the many farmers who have employed consultants over long periods suggesting that they at least believe they are getting value for money.

Part of the dynamism of objectives is the changing situation farmers find themselves in. Most farms follow a kind of life cycle in that in their youth farmers and their spouses are young and enthusiastic with values formed by their families and community. As the years pass the presence of children in the family often impact on values, and then eventually when they leave and the years start their toll, values and objectives undertake further change. Thus, what is right for one family will be wrong for another depending on their stage in life. With time farmers tend to become more conservative and change less. And where a farm has been in the family for several generations, the objective set is probably altered for the attachment to the land can be powerful leading to sustainability and environmental issues that will often be stronger relative to a farmer on a new farm.

Values are partly determined by a farmer's personality so given that it can change in earlier life, values can similarly change. Other factors giving rise to change are extended family influences such as, say, a financial input by relatives to enable expansion, a change in ownership structure (perhaps part of the transfer of the assets from one generation to the next). Similarly changes in the general farming environment may lead to changes. If there is a major shift in, say, prices leading to the impossibility of paying back a family mortgage it will mean objectives have to change. When an objective cannot be achieved the human condition usually leads to change creating objectives that are more realistic. If modifications did not occur, emotional turmoil in seeking the impossible could easily ensue.

Finally, it is important to accept that farmers are seldom totally stable in expressing their values and objectives. Over, say, 2 years a farmer might well make a rational decision over fertilizer application in the spring this year, but be quite irrational next spring. The major reason, all other things being equal such as prices and costs, values and objectives, is the impact of emotions. Being human means we often react according to our emotional state. Perhaps the local farmers' organization has voted the farmer off the committee for no apparent reason other than the words of a rival making the farmer angry at his neighbours. In an irrational response he orders and applies too much fertilizer going well past the point of marginal cost equalling marginal return in an effort to show up his neighbours as bad farmers. Farmers are not machines and, like all humans, react to their emotions. We cannot always help it. Atkinson *et al.* (1996) talk about the components of an emotion as being the following:

1. A subjective experience.
2. Internal bodily responses, e.g. a fright response.
3. Cognition about the emotions and associated situation.
4. Facial expression (you have all noted embarrassment).
5. Reactions to the emotion.
6. Action tendencies.

Of course, the good manager takes time to review an action relative to the objectives, and hopefully controls irrational action. Some people need help in maintaining this rationality.

Quantifying the Origins of Objectives

Introduction

To understand the origins of a set of objectives, and therefore consider how to possibly modify a farmer's, and a farm family's objectives, studying the factors influencing the objective set is important. This requires data. For students interested in researching objectives they should consider Garforth and Rehman's (2005) review of values, goals and objectives. In this they consider the alternative data collection methods. They all involve questioning farmers in various ways.

Presented below is an analysis of data collected through a postal survey of over 2000 farmers yielding 735 responses. While the data were collected to quantify the important factors in determining managerial skill (Nuthall, 2009), much of the data were also useful for exploring the factors leading to specific objectives.

The important factors

Listed below are the factors which logic would suggest are likely to determine objectives. How such factors influence the combined family objective set is another matter which may or may not be a simple amalgam of each individual's objectives. This is somewhat of an unknown in a research sense. The relative dominance and importance of each individual will impact on the group objectives, which may well be somewhat fluid. Too much dominance can, and does, lead to resentment which eventually surfaces in a range of ways.

For an individual, while each case will be unique, the likely factors influencing objectives are:

1. The parents', and possibly the grandparents' objectives and outlook on life.

Along with most factors the childhood environment will have a major influence on objectives. And similarly the genes inherited if in fact a person's genetic makeup is important to the values and objectives. As part of the family influence are the family traditions (e.g. perhaps employing rather more of the local people than might be prudent as they have always employed from specific local families; being heavily involved in the local church and its values). Also an important part of the parental influence will be the encouragement and general support both for school and other activities.

2. The farmer's management style (personality) and intelligence, which are also, of course, caused, in part, by the genetic heritage.

A person's ability to think through issues resulting from his or her intelligence will influence the objectives. For example, reasoning relating to management systems and their impact on the environment will affect both the view on acceptable systems, and the goals resulting from the reasoning. The same comments can be applied to personality, so, for example, someone with high anxiety levels is likely to have a stable income as an objective in contrast to a maximizing income.

3. The farmer's age, sex and education will influence the objectives.

Age clearly is part of the life cycle and also age itself can alter how a person sees his or her place in the world, and certainly the farmer's gender in that the hormonal system influences how people view life and what is important. Education to higher levels can also influence how people think about their existence and what might be crucial outcomes in their life. The type of school(s) attended, and the specific mix of teachers experienced can have a powerful influence, including whether the schools were rural- or urban-based.

4. The resource situation a farmer finds himself in will also impact on the objectives if for no other reason than they determine to a certain extent what is physically possible.

Often people adjust their goals to something which is practical and feasible. But on the other hand, some people are very determined to achieve certain goals and set their sights very high, and a subsection of such people will make it happen (and no doubt some have very good luck helping the cause, such as a long run of favourable seasons and prices). The resources available include the quality and size of the farm and its attributes, and the lack of a high debt load. The number and quality of the employees also impact on what is possible.

5. In a similar vein, the farmer's inherent managerial skill can impact on the objective set for it will influence both financial and other success and, therefore, what is practically feasible.

Associated with managerial ability will be the experiences a farmer has had, and the early life and encouragement offered by his or her parents and so on as noted in Chapter 3.

6. A farmer's inherent enjoyment of working with soils, plants and animals can be a powerful factor in objectives and in many cases is the reason he or she will accept a lower income relative to what might be possible from selling up and working in a city.

Similarly their interest in the natural environment (evidenced by the many farmers who have fenced off areas to allow indigenous species to flourish) and sustainable production that will improve rather than deplete the environment are all factors in setting objectives and goals.

7. And finally, a farmer's interest in people per se.

This factor will be partly determined by the farmer's personality, and the family background, but it needs mentioning separately for some farmers will put special effort into helping others and their community as a specific objective. An example would be the farmer who devotes many hours to farmer politics and lobbying politicians to enhance the farmers' general cause. This usually comes at the cost of income and an enjoyment of farming for its own sake.

Some quantification

The data from the farmer survey were used to develop linear equations to reflect the importance of each variable for which data were available. This was carried out for each factor, or component, of the farmers' objective set. In the surveys reported in Chapter 3 five distinct factors were numerically isolated. In this survey, these same factors emerged from the factor analysis, but also a sixth factor was apparent. Somewhat by chance, the sample had in it managers who were somewhat reluctant farmers. These were people who did not enjoy farming, but had no other qualifications or opportunities enabling them to leave farming (or perhaps some were rather timid). This list of objective components is as follows:

- Work enjoyment   To simply enjoy the process and environment,
- Wealth   To obtain wealth through both income and asset value increase,
- People   To enjoy contact with people including farm workers, other farmers, and family including being able to set up children with assets,
- Leisure   Organizes farming to ensure sufficient leisure and holidays, community activities and time to try new things,
- Risk   Uses farming systems that ensure risk levels are kept low,
- Stoicism   Continues farming as an objective simply as he or she has no alternatives.

The totality of any farmer's objective set is made up of proportions of these components, each of which has a different degree of importance for any particular individual.

The variables that entered the equations were (unless otherwise stated most variables took on a 1–5 scale reflecting the degree of truth in the statement) the following:

Age   The age, in years, of the farmer
Ed   Highest level of education (1 = primary, …, 5 = 3 or more years tertiary)
Lab   The number of employees on a farm
Gen   The gender of the farmer (1 = male, 2 = female)
Siz   A measure of the economic size of the farm being the product of the area (ha) and a code reflecting the intensity of the farm type (6 = intensive horticulture, …, 1 = extensive sheep)
St1   Management style factor: concern for correctness (anxiety)
St2   Management style factor: thoughtful creator (openness)
St3   Management style factor: conscientious planner (conscientiousness)
St4   Management style factor: community spirit (extroversion – community)
St5   Management style factor: consultative logician (extroversion – family and friends)

St6   Management style factor: benign manager (agreeableness)
Ee1   Childhood early experiences: helped with farm jobs
Ee2   Childhood early experiences: reason for decisions explained
Ee3   Childhood early experiences: listened to discussions on finances
Ee4   Childhood early experiences listened to discussions on technical matters
Ee5   Childhood early experiences: asked my opinion on decisions
Ee6   Childhood early experiences: wanted to know reasons for decisions
Es1   As a child encouraged to use imagination to find solutions and methods
Es2   As a child encouraged to improve observational skills
Es3   As a child encouraged to 'get along' with friends and relatives
Es4   As a child encouraged to learn agricultural knowledge
Gen   Number of generations on the particular farm
PO   Parents' objectives are different
GO   Grandparents' objectives are different
SM   Self-scored managerial ability (1 = poor, ..., 10 = excellent)
EM   Managerial ability calculated from the productivity and other data (%)

The equations were calculated using all these variables together with a number of other similar ones. They were all the data available on the variables logically expected to influence a farmer's objectives. Given the results, variables that had a probability of greater than approximately 20% of being no different to zero were dropped and the equations recalculated. These are presented below with the significance figures of the remaining variables given in brackets below the coefficient, and the standardized coefficient in the square brackets. These standardized figures allow a comparison between variables to assess the importance of each as they neutralize the differences in the units of measure. The equations were all highly significant, and the percentage of the variance explained by the equations ranged from 10% (stoicism) to 31% (risk). These figures are not particularly high, indicating other factors are also involved including the variables listed as potentially important but for which no data were held. Clearly further research needs to explore details of the parent's influence, and similarly the impact of culture. Questions relating to these variables were not available in the questionnaire. Details of the equations are as follows:

$$\text{Work enjoyment} = 0.95 - 0.06\text{Age} - 0.76\text{Gen} + 0.08\text{St2} + 0.18\text{St3}$$
$$(0.02) \quad (0.13) \qquad (0.01) \qquad (0.07) \qquad (0.0)$$
$$[0.06] \qquad [0.1] \qquad [0.08] \qquad [0.17]$$

$$+ \ 0.1\text{Ee1} + 0.14\text{Es3}$$
$$(0.04) \qquad (0.0)$$
$$[0.08] \qquad [0.14]$$

The most important variables are age and the constant (which could well be explained by some of the unavailable variables), and this is followed by conscientiousness (St3), early encouragement to get along with people (Ee3), and then gender. This suggests personality and parental influences give rise to work enjoyment.

Wealth = 0.29 − 0.05Lab + 0.09St1 + 0.16St2 + 0.13St3 − 0.9St6
(0.45)   (0.08)   (0.04)   (0.0)   (0.0)   (0.03)
              [0.07]   [0.09]   [0.17]   [0.13]   [0.09]

− 0.06Es3 + 0.05Es4 + 0.12Ee5 − 0.06Ee6
(0.14)   (0.15)   (0.0)   (0.15)
[0.06]   [0.06]   [0.14]   [0.07]

+ 0.05PO − 0.18SM
(0.11)   (0.02)
[0.07]   [0.11]

The source of the wealth objective is rather more complicated than 'work enjoyment'. The constant is not nearly as important, and furthermore, with a significance of 0.45 the chance of it being no different from zero is high. The three most important variables are openness (thoughtful creator) and conscientiousness from the personality variables, and the parental influence of encouragement to be involved in decision making at an early age (Ee5). The other variables also help determine the importance of the wealth objective, but are less important. The desire for wealth clearly has many components.

People = 0.1 + 0.05Lab + 0.05Age − 0.09Ed + 0.3St2 + 0.23St4
(0.78) (0.05)   (0.16)   (0.02)   (0.0)   (0.0)
            [0.07]   [0.06]   [0.1]   [0.3]   [0.22]

+ 0.18St5 + 0.06Es3 −0.04Es4 − 0.05Ee2
(0.0)   (0.12)   (0.17)   (0.15)
[0.18]   [0.06]   [0.06]   [0.06]

+ 0.06Ee6 + 0.06GO
(0.09)   (0.03)
[0.09]   [0.08]

The desire to interact with people as much as possible similarly has many constituents. The personality variables again feature (St) with openness and extroversion being particularly featured (St2, St4 and St5) as would be expected. Education also impacts through, perhaps, giving confidence.

Leisure = 0.49 − 0.12Age + 0.11St1 + 0.13St3 + 0.25St4
(0.01)   (0.0)   (0.0)   (0.0)   (0.0)
              [0.12]   [0.12]   [0.13]   [0.25]

+ 0.15Ee1 − 0.06GO
(0.0)   (0.03)
[0.12]   [0.09]

Once again the personality variables are strongly linked to the desire for leisure (particularly extroversion which perhaps expresses a need to take time off to interact with people in spare time and holidays), but also age with older people being less interested in leisure.

$$Risk = -0.98 - 0.2Age + 0.11Lab + 0.3St3 + 0.07St4 + 0.03EM$$
$$(0.0) \quad (0.0) \quad\quad (0.0) \quad\quad (0.0) \quad\quad (0.06) \quad\quad (0.0)$$
$$[0.21] \quad\quad [0.17] \quad\quad [0.3] \quad\quad [0.07] \quad\quad [0.34]$$

The large negative constant suggests people are inherently risk preferers, but with this being strongly negated by conscientiousness (St3) and age. Thus, as farmers get older they tend to get more conservative as is often observed in the field.

$$Stoicism = -0.82 + 0.09St1 - 0.16St2 - 0.06St3 + 0.01EM$$
$$(0.0) \quad (0.03) \quad\quad (0.0) \quad\quad (0.1) \quad\quad (0.0)$$
$$[0.09] \quad\quad [0.16] \quad\quad [0.06] \quad\quad [0.17]$$

This rather less important objective, which is more about an accident of history than an objective, is largely explained by a farmer's personality with openness being particularly important as is managerial ability (EM). Farmers with a reasonable ability stick with farming despite not enjoying the occupation, and farmers with a high level of openness, on the other hand, tend not to be stoic and probably tend towards giving up farming.

Overall, while quite a few variables clearly give rise to the objectives held by farmers, it is clear that their management style (personality) is a major factor. This set of variables is conferred on a farmer by the farmer's parents and early environment. Age, education and farm size also have some influence as you would expect. Finally some of the early experience variables ('helped with farm jobs' which implies the farmer lived on a farm from an early age, 'decisions explained', 'party to discussions on family finances', 'enquired about the reasoning behind decisions' (again implying early age farm experience)) were also significant.

Thus, to change a farmer's objective set involves working with the farmer over changing his or her personality, and similarly the impact of some of the early experiences. This raises questions of what is ethical. Many would conclude that an advisor or consultant is empowered to work with a farmer to improve the attainment of the stated objectives, not to change the importance of each objective. Indeed, in the early days of consultants working with farmers many were later criticized by the farmers for pushing them into intensive farming with a view to improving profit. This often worked for a while but when conditions altered, considerable stress resulted from both disastrous downturns and sleepless nights from the high risk, let alone the family reaction to less leisure time. More latterly, advisors and consultants accept farmer's objectives for what they are rather than assuming everyone wishes to simply maximize profit, or have the same objectives as the consultant.

Many would consider an ethically acceptable approach is for a consultant to work with a farmer, and his family, in reviewing their objectives to see if they are indeed what they want. It is also ethical to help a farmer whose emotions often override sensible decisions to maintain a more rational approach when making decisions aimed at achieving their objectives. This support requires training the farmer to consider decisions calmly and rationally in contrast to 'spur of the moment' reactions. This will likely require constant contact between the people involved as decisions are made in the initial phases or training.

The action steps to be followed in reviewing a farmer's objectives are as follows:

1. Set out what the farmer and family believe are the objectives. This will involve:

    a. Use of the questionnaire on objectives.

    b. Observation of the actions taken by the farmer.

    c. Talking to the farmer and family and asking them to come to an agreement on the objective list through reviewing both the questionnaire results and the observations.

2. Discuss with the farmer and family whether they believe the objective list and their priorities are in fact what they want. The process followed in (1) will frequently start the participants thinking whether in fact their list is appropriate, especially as it is unlikely they have ever before enunciated what they want from their farm.

3. Review whether the revised list is actually what the farmer and family want after allowing time for the participants to mull over the conclusions. Often this should be a constant process as conditions and life cycle stages change. Objectives are dynamic.

4. Make plans and organize support systems appropriate to the revised objectives.

Locus of control

As noted in the introduction, a farmer's LOC may be hindering achieving the objectives and so working with the farmer to help develop a more appropriate locus can be beneficial. It was noted in Chapter 3 that a farmer's locus was heavily dependent on his management style. But, similar to the objectives, the origins of his beliefs may involve further variables. To test this idea, the same set of data used for the study of objectives was used to develop an equation. The equation with only the significant variables included was as follows:

$$\text{Locus of control (\%)} = 52.98 + 0.85\text{Ed} + 1.53\text{St1} - 2.63\text{St2}$$
$$(0.0)(0.0)(0.0)(0.0)$$
$$[0.11][0.19][0.32]$$

$$+ \ 0.7\text{St3} - 0.51\text{St4} + 0.41\text{St5} + 0.41\text{Es4}$$
$$(0.01)(0.09)(0.19)(0.08)$$
$$[0.09][0.06][0.05][0.06]$$

$$+ \ 0.81\text{Mis} + 0.12\text{EM}$$
$$(0.0)(0.0)$$
$$[0.13][0.19]$$

where the variable names are the same as used in the objective equations with the addition of Mis which is a 1–5 score on how quickly the farmer takes up a lesson from experiences. The equation was highly significant, as were most of the variables. The equation explained 32% of the variance showing that other factors are also involved.

It is again clear that a farmer's management style (the Stx variables) is very important in creating his locus, particularly the openness character (St2), but also education, an ability to quickly learn from experience (Mis), and managerial ability. The latter is somewhat obvious and unhelpful, whereas working on improving the uptake of the lessons from experience, and similarly the openness character in management style, may well lead to a more appropriate locus.

Entrepreneurship

Some farmers aspire to be an entrepreneur; others are content to follow the well-proven paths of conventional management systems using what productive resources they currently command. However, it is useful to comment on the factors associated with entrepreneurship for some farmers may benefit from reviewing their objectives relative to being more adventurous. Clearly, objectives and entrepreneurship are highly related in that if a farmer has maximum profit as a high priority then acquiring the characteristics of entrepreneurs may well be beneficial. But can they be achieved?

A search of the literature shows that while entrepreneurship is widely recognized as an important phenomenon, there is a paucity of work explaining and proving it can be fostered. However, the research does show that entrepreneurship is related to a number of personal characteristics where it is assumed that entrepreneurship is reflected by business success.

It is commonly suggested that positively correlated to entrepreneurship are the following:

- the need for achievement;
- power; and
- leadership.

Also important is the degree of personal ownership of the business with the higher the proportion, the greater the entrepreneurship. In agriculture, of course, the manager's ownership level is frequently very high. The tests used in assessing entrepreneurs have, in addition, shown a positive correlation with:

- the need for influence;
- verbal aggression; and
- hostility.

Stewart et al. (1998) showed that entrepreneurs were higher in achievement motivation, risk-taking propensity and preferences for innovation. They talked about the profile of an entrepreneur as being a driven, creative risk taker who plans heavily. Clearly they have goals of profit and growth. What is noteworthy is that risk-taking is heavily linked to the anxiety personality trait and this trait can be influenced through proper training and support. To this end entrepreneurship can be fostered.

In an extensive review, Judge and Ilies (2002) also conclude that performance motivation is highly correlated with anxiety and conscientiousness, and that the other personality traits (openness, extraversion and agreeableness) are also correlated, but not so strongly. Taken together the five traits were 49% correlated with the motivational criteria they used.

In a specifically agriculture-directed piece of research (de Lauwere, 2005), it was concluded that all of the following affected entrepreneurship positively, but that 'love of ease' and passivity had negative impacts:

- self-criticism;
- leadership;
- creativity;
- perseverance; and
- initiative.

Overall, like motivation, entrepreneurship is clearly highly related to management style (personality) with the farmer's goals also having an influence. A farmer interested in, say, sustainability, in contrast to maximum shorter-term profit, can have all the right personality traits, but would not be labelled an entrepreneur in the conventional sense. For a farmer reviewing his goals and methods, he might well consider counselling and support to modify his personality traits if he concludes he wishes to be more entrepreneurial.

For the case study farmers, both regard themselves as entrepreneurs, though Hank has questioned his approach more latterly in this respect. On further reflection Hank did concede 'compared with most farmers I guess I'm quite entrepreneurial'. He is certainly that for he has grown the business in leaps and bounds over the years and now has two separate dairy farms which are both growing in output. When asked the reasons for his entrepreneurial approach Hank was somewhat lost for words. He did note, however, that a favourite Uncle probably influenced him with his stories of expansion and business growth. He aspired to follow in these footsteps.

Margrave comments that out of necessity he became an entrepreneur. This might be said about all entrepreneurs in that they either have to be venturesome to survive, or their management style and objectives drive them to follow the entrepreneurship path. When first taking over the family farm, Margrave experienced a major downturn in the agricultural 'climate'. Subsidies, incentives and agricultural protectionism were all suddenly removed by the Government at the same time as prices were weak. The upshot was he had to branch out into all sorts of additional activities to be economically viable. Examples include rabbit farming and wood supply activities. In the end, none of these initiatives was sufficient and the farm was sold to eventually be replaced by other arrangements. Despite this, Margrave believes he is an inherent entrepreneur through his keenness to try new things: 'makes life interesting, gives life an edge...I always enjoyed walking out of the house to start work each day and have this very high motivation. Part of my personality I guess'. Yet his father was not an entrepreneur, perhaps because he went through the Great Depression and this created a cautious approach to debt and unproven ventures: 'my father even straightened out old nails to save expense'.

Concluding Comments

Anyone studying the human factor in farm management must recognize that the farmer and family must base their decisions on a set of objectives which accurately reflect their wants and needs. Objectives are very much part of the human side of farming and these involve considerably more than the simple material outputs including cash profit. The classical assumption of maximizing profit is a total misnomer except, perhaps, where the owners of the business are totally divorced from the physical location and operational management of the farm.

Assessing the objectives is a human problem, as is checking that the professed set is correct. In that farms worldwide are largely family-owned and family-managed, assessing objectives must involve the whole family. Each family will be unique in its makeup and objectives, and in its cohesion. A successful family will consider each member's wants and needs, and communicate successfully leading to democratic conclusions accepted by all.

Clearly there is competition for resources given the farm and family are intimately related so resources devoted to the farm means the family will need to go without, and vice versa. However, some outputs from the farm will directly provide family satisfaction thus representing a complimentary situation. An example is the space, views and peace provided.

One of the first things an observer must note are the 'farm's' objectives for it is not possible to judge success without this yardstick, and therefore what might be improved. Part of judging the situation is noting the family cohesion with particular emphasis on the relationship between spouses. Where disharmony exists, everyone can gain if this can be approached and lessened, perhaps with outside professional help.

Farming involves many stresses and in the bad patches disharmony can have a major impact on efficient management. In a bad summer, for example, where the farmer is working very long hours without family support a break point can be reached. Furthermore, each family will go through a farm life cycle with some of the stages being potentially more stressful than others. When the children need secondary or tertiary education the stress on available funds will be extreme, especially if this occurs in a low point in the inevitable price cycle. Education may well involve the children living away from home thus further creating stress and a family dynamic that is totally changed with just the farmer and spouse being left at home with, perhaps, time on their hands.

In most cases a spouse will be female, but whatever the case, 'marital' harmony is an important aspect of efficient resource use, particularly where the spouse is heavily involved in both working on the farm and in decision making. An observer must talk to both partners and assess whether improvements are possible, and call in what help is regarded as necessary provided the request comes from the spouses. The observer needs to be able to move the participants to the point where they realize there is a problem that might be retrievable. Or perhaps enlist the help of others in the family or community. Assessing whether improvement is possible partly involves assessing the personality of the participants for the available research clearly shows the link between personality and marital success.

Also important are the basic values and backgrounds of the spouses. It was clearly noted that core values lead to objectives which in turn create goals to be achieved. If spouses have quite different values it is hard to achieve cohesion unless they have benign personalities that enable major compromises.

Another important aspect of personality is how a person reacts to emotions. Being human, farmers have emotions that sometimes override rationality and so it is important for a farmer to recognize this and take time to consider particularly important decisions. A stable and helpful family situation helps attain emotional stability and appropriate decisions. Human beings can get depressed, and some extremely so. But family support, and professional help in some cases, can negate this depression and aid rational decision making. The state of any depression is not simply black or white, but rather a condition that everyone suffers at some stage. There is a continuum from euphoria to the deepest depression but with most having simple 'down periods'. In that farmers are dealing with many uncontrollable factors, if they all turn down at the same time anyone would get somewhat 'down' in their feelings. Major problems can occur, however, when both the weather and markets are unfavourable for a long period.

Finally, if an observer understands why a particular set of values, objectives and goals are held it is much easier to work with the farmer and family in checking their appropriateness and possible modification. The research available suggests personality has a major influence on objectives, and this in turn is dependent on parents, the early environment and community. Assuming a family wishes to check, and possibly change their objectives, and perhaps modify their behaviour to provide better family harmony and, thus, more efficient decision making, consideration must be given to the ethics of influencing them using a range of methods. The key is to let the family drive the operation following totally open discussions and conclusions. Better attainment of the truthfully held objectives is the goal, and where attained, provides considerable satisfaction to all involved.

8   Methods of Improving Managerial Ability

Introduction

The main reason for studying managerial ability is to consider ways of improving the farmer's managerial skill, though an understanding can also be useful when considering the impact of agricultural policy initiatives. The purpose of this chapter is, therefore, to consider the techniques that will improve a manager's skill no matter at what level they start.

As every manager currently exhibits a particular level of ability, a set of methods that can initiate improvement in all situations is required, and is highly desirable. Some farmers will improve more than others both due to their starting point and inherent ability. Each starts with a certain potential as defined by their genotype, and their early environment and experiences. While the genotype is fixed, additional training of various kinds can change and improve the impact of their experiences. Fortunate farmers will have an appropriate genotype (intelligence, personality, etc.), and appropriate early experiences in the form of family life, education, challenging situations, encouragement and training courses. These all lead to skill, curiosity, confidence and self-esteem. Farmers without these advantages must work at compensating their situation with the support of all the facilities that are available.

Deciding on a manager's ability depends on the use of benchmarks including technical and economic standards. The latter are easy to obtain and calculate. What is not so easy is the testing of a farmer's aptitude and managerial style with particular reference to their personality. While general tests exist, tests for attributes like self-esteem, motivation and people skills specifically for farmers are not yet available. Thus, an assessment is dependent on observations of a farmer at work. Such assessments rely on the concurrent observation of other farmers to help form benchmarks of what is possible. For technical and financial figures, various techniques are available including data envelope and stochastic frontier analysis in which farm data are used to create the most efficient boundary of production against which others can be compared. In any improvement programme, constant reassessment is required to gauge what changes are being achieved.

It is an inherent assumption that farmers can change their ways, if their desire is sufficiently strong. This relates to the argument on 'plaster or plasticity', referred to earlier, which looks at whether people are fixed in 'plaster', or whether change is possible. The research makes it clear that 'plasticity' is the reality.

The earlier chapters have covered many of the aspects that make up a farmer, and thus refer to what might be changed with benefit. The following were noted about recognized experts:

- They were excellent in a limited domain (area of expertise).
- They were good at defining a problem (anything that requires a decision).
- They can accurately observe the relevant information.
- They can perceive meaningful patterns (structure of a problem).
- They have superior short- and long-term memories.
- They come up with the best solution.
- They have good self-monitoring skills (improve their expertise).
- They spend time on sorting problems they have not previously experienced, and from this create and improve intuition (tacit knowledge).

Thus, good managers will have all these attributes. In addition, it has been pointed out that farmers need to:

- know the technology;
- be good at recording the relevant items they have observed;
- be good at anticipation and planning;
- have a good knowledge of decision rules and principles;
- have excellent people skills;
- have an appropriate personality, or management style; and
- have appropriate intelligence.

This is clearly a demanding list.

Intelligence wise, a mind able to think logically is important, as is one that can hold and retrieve the relevant information as required. Logical thinking is related to learning from experience which involves self-criticism and analysis. Whatever the objectives, the farmer's attitude to risk impacts on what actions are optimal, but extreme attitudes are probably not helpful. And the farmer's belief in what control he does have (locus of control) is relevant in that it might be different from reality and therefore impeding progress. Risk and control attitudes are more than likely correlated with a farmer's inherent management style, or personality.

The five factors in personality impact on a farmer's style, which in turn impacts on likely managerial success. Conscientiousness is obviously important, as is 'openness' for it confers original thinking and a spirit of 'trying out the new'. In order to develop good relationships, a degree of extroversion is beneficial, as is agreeableness, for this confers a benign approach to others. A certain amount of 'anxiety' is valuable in that it stimulates timely action and care, but too much causes stress and subsequent irrationality. In efforts to improve management, thought must be given to changing the expression of these personality characteristics.

It was pointed out earlier that intelligence is partly genetic and partly developed (crystalized). It does appear, therefore, that intelligence can be enhanced with appropriate training. Certainly memory can be improved as can creativity (an important aspect of coming up with solutions), logic, learning from experience, and spatial aspects such as understanding maps, perspective and all tasks requiring spatial organization. Much of the mind's work relies on pattern matching, and so developing images, patterns and their efficient retrieval is an important function of the mind.

The model of ability discussed in an earlier chapter brings together the major factors reviewed above. Besides style and intelligence, it is clear that experience is an integral part of the evolvement of a farmer's ability. Any management skill improvement programme must look at how the lessons of experience can be assessed and better utilized.

Reasonable intelligence impinges on learning from experience through problem recognition and the use of benchmarks. The ability to form appropriate chunks in any problem is also important, especially in primary production with its biological connotations and consequent interrelationships. An appropriate intelligence also enables decisions which allow for the dynamic nature of production. Effectively, there are interrelationships across time which must be understood and integrated into decision making. Further, management complexity comes about through risk and uncertainty, the impacts of which must be clearly understood, as, if nothing else, primary production is certainly full of risks. A flexible management style is necessary as initially made plans may need changing as production unfolds. A person who cannot readily change and think 'on his feet' will find success difficult.

A reasonable IQ is also important in understanding the decision rules that lead to optimal production. For efficient resource allocation, a farmer should follow the rules derived from a study of production economics, and also of similar economic theories such as cost–benefit analysis. Fortunately there are many texts devoted to such subjects, though few farmers formally study them leaving their logical skills to lead them to the lessons such as the 'marginal return equals marginal cost' rule.

An earlier chapter considered the specific competencies farmers believe were important. Clearly any training programme must take this list into account. It will be recalled that the successful general competencies heading the list were:

- risk management;
- observation;
- anticipation;
- negotiation;
- planning and the associated analytical skills;
- learning from experience;
- skill in dealing with people;
- successful implementation of plans;
- an excellent knowledge of the relevant technology; and
- an ability to integrate solutions across all components.

Also regarded as being important is an ability to achieve:

- timely action following timely observation of the signs that indicate action is needed.

Furthermore, part of fulfilling the competency list is having good skills for:

- watching;
- reading; and
- listening.

The challenging lists continue.

An earlier chapter highlighted the common decision biases, and noted that stress can give rise to irrational decisions, and the last chapter mentioned that emotional factors can, and do, influence decisions. While methods of stress alleviation were discussed, there is no simple answer to overcoming bias. In a general sense, any managerial process that gives other than optimal decisions is biased. Overall, then, the earlier chapters have covered the many things that make up a manager and his skill. The essence of a plan to reduce decision bias and improve managerial skill is to tease out the core factors and institute a programme of correction where deviations exist. While most farmers are either consciously, or unconsciously, working in this direction, it is important to consider the kit bag of tools available to assist this process. The rest of this chapter moves in this direction by providing ideas and concepts on the processes of learning, the kinds of benefits possible from a good programme, what topics should be tackled, and the ideas on the resources available for helping, and, finally, a list of the training methods that might be considered. The chapter is completed with some concluding comments on the book.

Learning Processes

Everyone learns best in slightly different ways. You have already been introduced to Kolb's test which lets you conclude whether a person is a 'concrete' thinker who best learns from practical situations, or at the other extreme, is capable of learning from abstract resources such as a text book. Whatever the case, it is fair to note that most people learn from practical exposures, though this may not be the most efficient in that a computer screen, or book, is a cheap resource, if in fact a farmer can obtain meaningful skills and knowledge this way. But few can.

First, learning will seldom occur, if the person is not a willing student. An open mind that is keen to change and absorb any lessons on offer is a good start. This is not to say that someone with opposite inclinations will not learn and change their views. Clearly, some bad mistakes that are financially disastrous will most likely lead to a change of views and improved management, provided the person can work out what went wrong and why.

Moving a farmer into an appropriate frame of mind can be achieved by providing information that shows improvement is possible. This information might be comparative farm statistics ranking the particular farm somewhat down the list. Farmers who are not over confident are clearly going to be more receptive. Furthermore, if what is being learnt is exciting there is a better concentration on the process. Ewel (1997, p. 9) comments 'learning occurs best in a cultural and interpersonal context that supplies a great deal of enjoyable interaction and considerable levels of individual personal support.'

Second, the farmer must be exposed to improved methods, knowledge, ideas or whatever is appropriate for his situation. This means pointing out the process currently used by the farmer, and comparing this with an improved approach. A simple example would be methods of creating a whole farm budget. In some cases, perhaps the farmer has never created a budget, and so the training starts with a blank slate, though it would be surprising to find farmers who have never produced some kind of rough budget, even if perhaps mentally.

Practice is then required to reinforce the correct method and ensure the process has been learnt. This step may take some time, but eventually will be achieved. Think back to learning to ride a bicycle. Initially it was a difficult and painful experience, but eventually the process became in-built and automatic no longer requiring obvious conscious thought. There is no reason why this process should not equally apply to all the skills required, though some will take longer than others. An example of the latter might be forecasting international prices of a commodity.

Research on children (Nuthall *et al.*, 1993; Nuthall, 2007) has concluded that repetition is important in that there must be a number of instances, and types, of relevant experiences for the student to develop a construct or understanding of an idea. The time interval between experiences is also important. It also seems that much of the learning occurs from within student group interactions and discussions in contrast to the formal lessons. Most people would recognize this situation in that seldom do you comprehend a concept or idea without several exposures and reinforcement. When it comes to simply remembering a fact this is probably different. If someone tells you that it is necessary for you to pay your tax bill by, say, the end of July, most people will only need to be told once!

In a similar vein, Enos *et al.* (2003) believe 70% of learning relates to informal learning leaving 30% occurring through formal training. They comment (p. 379): 'Informal learning is a continuous cycle of challenging experiences, action and reflection. Informal learning for managers is a social process (social interaction with others in the workplace).' It is interesting that the situation seems to be identical no matter what the age of the learners.

There are many examples of this kind of conclusion. One other example (McCall *et al.*, 1988) found that 30 of the 35 managerial job skills they believed are important were learned informally, and Beaudry *et al.* (2005, p. 1) stated: 'The human resources literature suggests managerial skills are difficult to codify and learn formally, but instead tend to be learned on the job.'

All this research emphasizes the interactive and social aspects of learning; thus, sometimes learning can be assisted in the rural environment through farmers getting together in groups (Fig. 8.1). These groups can be formally organized, or perhaps might just be neighbours regularly contacting each other. Similarly the farm family might be involved.

Fig. 8.1. Farmers enjoy seeing the systems their colleagues are using. Such field-day comparisons stimulate an examination of the farmer's own system, and critical introspection.

The third important factor in learning is feedback. Having absorbed a lesson and implemented it (e.g. learning about and assessing the risk associated with winter stock feeding and subsequently ensuring certain feed reserves), observing the resultant outcomes clearly leads to improving the decision rules created. Effectively, as Kolb *et al.* (1974) noted 'learning is a continuous process of concrete experiences, reflective observation, abstract conceptualization, and active experimentation'. Some refer to this as 'experiential learning' (Kayes, 2002). Feedback might come from the farmer's own observations and records, or perhaps from consultants and colleagues. Positive reinforcement is always important.

Mention was made in an earlier chapter about Kelly's idea of 'man the scientist' in which he is always trying to come up with constructs (rules of thumb) that suit the evidence observed. Such constructs should be constantly under review to provide better systems that match the observed environment. Clearly, learning is a continuous process, and is what leads to tacit knowledge, that mysterious thing which drives us all, hopefully, in the right direction. A farmer's current knowledge and understanding will clearly influence the interpretation of outcomes and thus the lessons to be learnt. In a way, humans cannot prevent themselves from learning that most are striving to improve, it is just that some are better equipped to achieve this than others.

Self-'critiquing'is the fourth factor. The knowledge of one's strengths and weaknesses can be included in this aspect. It has been noted earlier that an important skill is being critical of everything observed, so that any information or idea must pass the farmer's internal test of reliability and accurateness before it is accepted. Similarly, being positively critical of one's analyses and constructs is important and is one of the processes involved in learning. This is part of the 'man the scientist' idea and leads to examining one's own processes leading to improved decision systems and knowledge.

Thus, in summary, a theory of learning that is widely accepted involves having a person who:

1. Is willing to change and has an open mind.
2. Can accept that improved ways do exist and is willing to search these out.
3. Can practise and re-practise a new approach until the lessons are learned (experiential learning).
4. Is given feedback, either self-created or from others.
5. Is self-critical.

As noted earlier, these are the attributes of experts and this is how they learn. Of course, other approaches do exist, such as reading a text book, but it is clear that farmers seldom use, nor benefit from, these formal approaches. The research does emphasize that managerial skills do need to be learnt on the job. If the objective was to become, say, an art historian, the situation could be quite different.

Consultant Bruce, being heavily involved in training farmers, has ideas on farmer's learning attributes. He is a strong believer that farmers are kinaes-thetic learners: 'Show me how and I can see what to do.' In a similar vein the 'trial and error' experiential approach is important: 'Give it a go and you will soon learn.' And Bruce comments that farmers learn a lot from their peers, particularly the respected farmers proving the point that if you can get the lead-ing farmers to change many others will follow. Overall, Bruce believes that a farmer must have an open mind: 'Hey, I've made a mistake, how can I learn from this?' The consultant can be an important catalyst in this learning process through being a mentor, and directing the farmer to role models.

Benefits of Changing

Most managers would like to improve their skills, and every manager can improve, if he or she has the desire. Clearly, the benefits involve increased profits and/or greater satisfaction with, possibly, greater resource efficiency. The amount of change will depend on the particular manager, and the resource devoted to the exercise. As with all investments, and a change pro-gramme is nothing less than an investment, the costs must be weighed against the benefits. Everyone embarking on a management improvement project must assess their situation in this respect. Eventually, however, improvements can become almost costless in that the right attitude of observation and change can become automatic, and the farmer becomes an expert with the constant review ability.

It is not possible to give figures on, say, profit increases relative to time invested, as each situation will be unique. But various comments of a general nature can be offered. Frequently, people running short courses for farmers seek feedback on whether the participants believed they gained. Virtually all published data shows a positive response from the farmers.

For example, Cameron and Chamala (2002) developed a number of tests which they used to test farmer's views on the value of an extension programme. The results compared pre- and post-training test scores and found considerable gains from the courses.

This study was conducted in Australia. Hanson et al. (2002) provide a US example which showed that training in financial and production management significantly altered farmer's views of various techniques with the farmers recording that they found the programme extremely valuable. Hopefully, the farmers then went away and implemented worthwhile changes to their management. More recently Jackson-Smith et al. (2004) found a link between a deeper understanding of financial concepts and greater returns following farmer's attendance at courses. The farmers also testified that they found the courses useful. The advantage of this latter study is that not only were the farmer's views assessed, but the profit outcomes were also recorded.

In a more general sense, over the years, there have been many studies on the value of education in the general population. Psacharopoulos and Petronis (2004) provide a review of many of these studies. For example, Ashenfelter and Rouse (1998) found there was a 9% increase in wages from each year of schooling using 700 identical twins to give comparisons. Furthermore they maintained that family background explains about 60% of the variance in school attainment. In an attempt to further quantify the relationship, Bowles et al. (2001) found that wages were given by the equation:

Wages = 0.196 years of schooling + 0.081 IQ
            + 0.035 years of experience
            + 0.025 parental socio-economic status (SES)
            − 0.096 number of children.

It is interesting to note the schooling and experience figures. Of course, this is for a sample of the general population, many of which would be employees, but nevertheless, the benefits of training and experience are strongly positive. In farming, it would be expected that experience would be more important.

When it comes to specific agricultural training, Kilpatrick (1997) notes that farmers with more education, and who attend more courses, have greater profitability than their peers. Kilpatrick (1999) also shows that the more the education, the greater the number of changes a farmer makes each year in his management, and the greater the number of training courses attended. It is also noted that US experience shows that 50% of the productivity changes made on a farm are due to learning on the job. The 'learning on the job' leads into tacit knowledge (intuition), which Sternberg et al. (1995) believe helps explain job success. They also note tacit knowledge was not correlated with IQ, and IQ explains about 25% of job success.

With respect to formal courses, both schooling and vocational short courses, it is clear that the overwhelming evidence points to very positive benefits, though it must be noted that the benefit to each farmer must depend on his unique situation. It must also be noted that for short courses, most attendees used in the research were probably farmers who were enthusiastic about learning. If the farmers had never attended a course before, and did not wish to attend but had somehow been persuaded to attend, the results might have been different. But you cannot say whether the outcomes would have not been good, and indeed perhaps the improvements might have been even greater?

In that considerable learning occurs 'on the job', the benefits from this probably relate to the help and guidance available. There is, however, no research quantifying the value of 'one on one' tutoring. What are available are the results from the extensive research on the benefits of personal counselling. Overwhelmingly the results are positive (e.g. Atkinson *et al.*, 1996), so you would imagine that the same would apply to similar management tutoring. Definitive research on this for farmers must await the future. What can be noted, however, is that in countries where private consultants are available, they are largely fully employed, and in cases where state-funded consultants are available, they too are generally very busy. Thus, the farmers must think that their help is valuable, though some of the assistance is rudimentary (such as making financial comparisons and yearly budgets) in contrast to working on basic management skills. However, not all farmers make use of the consultant's or extension officer's 'one-on-one' help. If more was available, it is not clear whether they would similarly be fully occupied.

Topics and Resources Available

Farmers have a wide array of resources available to help in improving their skills. Material available on the World Wide Web (WWW) can be accessed by large numbers, and certainly books can be accessed by all. In many regions, both long- and short-duration courses are available on a range of topics specifically for farmers. Some courses are run by state-funded extension organizations, and others are offered by non-government organizations, usually on a fee-paying basis. Some universities and other educational organizations have programmes accessible on the Internet, or by correspondence, and even attendance in person where this is practical with respect to the location of an organization.

While some farmers will be very clear on what they wish to learn, others will need assistance in selecting what to study, and in starting off. Even for the former, however, assistance can be useful for selecting resources that they will find useful and appropriately positioned. Farming colleagues can help, as can consultants and extension personnel. Farmers' organizations are sometimes similarly useful.

For planning skill improvement programmes, it is useful to have a list of topics that might be addressed. Planning, as the first skill, starts with the general skill of observation and understanding the facts; so, successful courses on observation might be a starting point. This is followed by the general skill of anticipation and all that entails, and then any programme would not be complete without studying risk and uncertainty.

Topics within observation which we will need to consider are:

- 'what' to observe;
- skills necessary for optical observation (scanning, interpretation, re-examination, review and repeat, snapshot storage, conclusion, notebook jotting);

- factors in reading (active reading, skim, summarize, question, recite and review, conclusion);
- listening skills (active approach with feedback and confirmation, attention, reception, perception, open mind, etc.);
- listing and prioritizing objectives (farmer, spouse, family, other owners, etc.);
- memory development (storage, retrieval, practice of repetitive association, remarkableness, visualization, etc.);
- problem definition;
- deciding relevance (critical view of proposals, offerings, etc.); and, finally
- record-keeping (what, how, legal requirements, cost and return of records, etc.).

No doubt you can think of other aspects of observation to be added to this list.

As the future is uncertain, the skill of making only decisions that must be implemented in the 'here and now' must be acquired. If made early, decisions that do not need immediate implementation may turn out to be suboptimal once the time for implementation arrives. The conditions are likely to have changed. Thus, while a farmer might have in mind what needs carrying out at some future date, the skill of having 'in mind' several possible actions is important, so that once the future becomes the present the decision suitable for the current conditions can be implemented. Of course, decisions should not be delayed beyond their 'use by date', for, in such cases, the operation will not be 'timely' and the outcomes less than perfect. It is a fine line. A farmer must learn an attitude that copes with delaying decisions to the last minute. Methods encouraging farmers to have this approach are clearly important and require the farmer to be 'clear headed and to hold their nerve'.

Anticipation must also include learning about:

- planning horizons (how far into the future is anticipation required to ensure the actions implemented now are optimal);
- strengths, weaknesses, opportunities and threats (SWOT) analyses to allow appropriate planning;
- developing imagination, creativity and visualization;
- ensuring practicality (realistic visions); and
- forecasting (reading, technology, markets, understanding relationships, both technical and economic, rules and regulations, etc.).

Imagination training involves:

- brain storming;
- use of random words for starters, and such techniques;
- logical thinking; and
- learning to break from traditional concepts and watching for emotional blocks.

Visualization is effectively a mind-based experimentation and involves pattern matching and so exercises to develop these skills are necessary. These mind experiments must allow for the correlations and interrelationships of the farm components; 'what impacts on what' must be a topic constantly practised.

Anticipation also involves putting plans on to paper and thus requires:

- budgeting skills (physical to financial to cash flow);
- critical path analysis;
- project management (monitoring and control, etc.);
- knowledge of all the regulations around change (employment contracts, resource consent applications, environmental impacts, etc.);
- time management; and
- feed budgeting where animals are an important component of the farm.

When it comes to implementing plans, it is important for the farmer to have the confidence to act on his convictions. This is a management style factor which needs consideration. Some farmers 'dither' and action suffers ending up in a 'muddling along' type situation. Thus, studying strategies to provide confidence and conviction is important to some (start in a small way perhaps, confer with supporters, etc.). Similarly, of course, the farmer's management style in general may need attention, but this 'topic' probably requires more than just a simple study of the procedures involved. The next section considers methods of achieving a change of a farmer's management style. The topic of 'people skills' also comes under this area. And the same comments apply to developing logical thinking and other components of intelligence.

Under risk management the topics that must be covered include:

- probability, chance and uncertainty;
- sources of risk (prices, weather, employees, etc.);
- tools to reduce risk and uncertainty (diversification, contracts, insurance, low-variability products, etc.); and
- calculations necessary to analyse alternative strategies.

One of the difficult topics in risk is understanding the idea that while you might make an optimal decision, in reality it turns out to be wrong relative to what you would have done given hindsight. Thus, the nature and features of risk and uncertainty must be studied. Similarly, determining a person's risk attitude must be covered, and how this can be used in planning.

For the resources available for each of the topics, comment has already been made about the Internet, and the availability of books. Similarly there are courses offered in many areas (e.g. one Australian group (Farm 500) has courses on financials, administration, risk, land, succession, farm safety, marketing, machinery, staff management, people skills, crop and stock production.) Books and articles abound on skills like negotiating, conflict management, assertiveness training, communication, listening and the like. Books on production economics as related to agricultural production abound, as do books on price determination and marketing.

Finally, it is important to mention decision support systems (DSS). This refers to farm computer-based packages (Fig.8.2), and also Internet-based, that provide training in management skills through their constant use. Any farmer using a good package will, depending on the system, gain skills in observing and recording relevant data and information, and analysing the data to provide

Fig. 8.2. In the early days all but the farsighted farmers scoffed at computers. But now they are an important part of their management team.

useful decision support. Packages should also relate to the farmer's goals and thus stress this aspect of being a good manager. An inquisitive farmer will want to know the assumptions inherently contained in the packages, and modify these to suit his situation where possible. Furthermore, integrated systems will stress just how the components of a farm are interrelated. Use of these packages will enhance a farmer's observation, anticipation and analytical skills. They should also allow for risk and uncertainty.

Training Methods

Overview of methods

Mention has already been made of how successful learning is most likely experientially based. The kinds of resources available to support any skill enhancement programme have similarly been listed. Skills that simply involve the use of listening, reading and watching can be picked up in a straightforward way by farmers, either on their own, or through some course or group activity. Examples are ideas on methods of reducing risk and uncertainty, understanding the basics of negotiation skills, and so on. For skills that involve other than simply learning the facts, experiential-type systems become important. This is where attitudes and tacit knowledge need enhancing.

Where an individual is working by himself on an improvement programme, and everyone should be doing this whether or not they are involved in any formal programme, the person must learn to constantly review what has happened with a critical view and tease out the lessons as a self-learning experience. Of course, few people actually live by themselves, so sharing some of this review and analysis with others locally based is nearly always valuable.

Where it is possible, formal assistance in reviewing the lessons of experience can be sought. This might be through extended family members, and/or compatible neighbours. It also might be through a group arrangement where the group sets itself up as a group mentor-type situation. Such groups might be organized through a farmer's organization, or perhaps just by locals who are keen to improve. Groups might seek professional help to guide their activities and offer ideas.

The other alternative is to seek 'one on one' professional help in making sense of the lessons to be had. The professional might be an agricultural consultant, or extension officer, or it might be someone who is skilled in decision making but does not necessarily understand agriculture.

This whole process of making use of experiential learning requires:

- noting carefully the sequence of events;
- the conditions that existed;
- the outcomes; and then
- reviewing them to see whether some other kind of response, or decision, to that taken might have been more appropriate.

Where others are involved, it is useful to retrace the farmer's thought process in coming to a conclusion and action.

In non-professionally supported groups, people take it in turns to expose what they have done to encourage both self-analysis and group input. Such groups are a regular feature of urban life, but often for different reasons (business mentoring and weight-watchers, for example). The group members might also quiz each other in turn for they may expose problem areas that the targeted farmer might not even recognize as a problem. These processes must, of course, be carried out sensitively. The members of a frank and open group must be compatible and possibly self-chosen. Furthermore, an absolute rule must be to retain complete confidentiality unless otherwise specified, for without this it is doubtful whether farmers will have the confidence to recount the full details of their thoughts and processes. Thus, one benefit of group member examination is creating better skills in problem recognition.

To stimulate situations that can be self, or group, assessed simulation games can sometimes be used, which attempt to create decision situations that allow a farmer to practise the skills acquired, and provide 'hindsight' information that can then be considered for providing lessons on what went wrong in the decision making. For simulation games to be useful they must be realistic and relevant to the environment under which the farmers are working. There is some evidence that games do have significant benefits in student learning (e.g. Neilsen, 1974; Menz and Longworth, 1976; Curland and Fawcett, 2001).

Mentoring approaches

Whether working on a 'one-to-one' basis, or in some form of support group, there are some principles and approaches to mentoring that have been developed in the world of counselling. These proven techniques will undoubtedly work when trying to improve the attitudes and management style of farmers. This is a separate problem compared to simply learning facts, figures and techniques. Generally the methods rely on:

- understanding why a problem exists; and then
- attempting to modify the current thoughts and behaviours that gave rise to the 'problem'.

The word 'problem' is used, of course, to represent any situation that needs changing. An example is where a manager tends to get over anxious. Another is where the manager gives little thought to the future prices and the collection of information that might help a farmer be better equipped to anticipate what might happen in the markets. In this example, there is usually plenty of help from the marketing experts who often write in popular journals. But the view of the farmer about what he should consider may need altering, and the farmer certainly has to look at the reports from an informed perspective to enable assessing their importance to his farm situation.

Management mentoring can be defined as a relationship between two or more people designed to improve the management skill of a farmer. For success such a relationship should be:

- non-judgemental;
- open and frank (for obvious reasons); and
- involve trust, sympathy and understanding.

The objective is to provide an environment and approach which ensures everyone is relaxed and confident, so that they can express their thoughts and ideas without any form of criticism.

Bruce the consultant is a strong believer in mentoring arrangements, but he similarly expresses the importance of the need to 'trust one another', and the need for 'accountability between each member of a group'. He also believes that a mentor group can find it very valuable to visit other regions and systems to get them asking 'why does that farmer do it that way?' This leads to the farmers questioning the reasons they follow in their particular systems, and perhaps together concluding that a change might be valuable.

Furthermore, learning about the background to a problem clearly helps others suggest alternative approaches that might solve the problem.

An example might be where a farmer tends to be very short-sighted when planning and making decisions. Perhaps he learnt his basic farming in a very reliable climate and needs to understand more about the current environment. Another example might be a farmer who has poor negotiation skills. Perhaps he is simply a strong introvert and is fearful of dealing firmly with people suggesting alternative solutions to whatever is being negotiated. Once the farmer understands his situation better, perhaps he and the helpers can develop strategies to overcome this situation. Maybe the farmer should never agree to anything straight off and then consult others before suggesting a change to the proposed deal. The objective is to give the farmer insight into why he has always fared poorly in negotiations, and develop an awareness of how negotiating might be better tilted in his favour.

Seldom is it possible to alter a farmer's attitude or approach overnight. It will normally be necessary to return to the 'problem' for reviews from time to time to constantly assess what progress has been made and reinforce the solution offered. Reinforcement is often useful in the form of some reward or other. Perhaps the farmer can reward himself with a half-day fishing, or golfing, if his mentor deems that progress has been made.

Another way to explore and reinforce appropriate management is to visit and discuss with people who are good at the skill under consideration. Farmers generally enjoy talking to other farmers about their methods, and so where a farmer is generally regarded as an expert, for example, dealing with risk, then his skills need tapping into. This method is called 'modelling', you are modelling off an expert farmer.

Then what is known as role playing can be useful. This is about practising an appropriate management skill. Perhaps a good farmer can act out how he deals with a conflict situation so that another farmer who often has difficulty with staff might then practise dealing with a difficult situation. One farmer acts out being an employee, and the other acts out being the farmer. Of course, before doing this they must have sorted out what they believe to be the methods that should be used in conflict management. The farmer recognized as being an expert is a starting resource, as are books on the subject. Where a 'one-on-one' situation exists, the consultant will have at least the theory of conflict management and so can role play the situation. Effectively, this approach is all about practising in a supporting and understanding environment. Assertiveness training might follow the same approach where a farmer needs to learn to be assertive, but not aggressive. Others might call it 'sticking to your guns'.

What is known as desensitization can sometimes be of benefit. This relates to situations that are of real concern to a farmer to the extent that he does not function efficiently. While this might be called stress, it usually refers to a particular situation causing concern, whereas for all other situations the farmer is totally relaxed. An example might be bidding at an auction for animals. Maybe in the past the farmer has made some bad decisions to the point where he now finds it difficult to operate rationally at auctions for fear of getting it wrong. The desensitization approach attempts to replace the concern with something incompatible with the anxiety. Perhaps simple physical relaxation will solve the problem. The farmer starts by imagining being at an auction and at the same time practises muscle relaxation starting with the foot muscles and moving up the body. In the first instance, a comfortable chair should be used! The next step might be some role play then eventually actually going to an auction that involves something minor and then slowly moving up to an important situation. This desensitization approach is called 'extinction'. You are trying to replace the problem with an alternative feeling so that the problem is extinguished.

Whilst working with others on perceived management problems is one option, there is nothing to stop a farmer from working on his own. Indeed, it would be surprising if most farmers did not, it is just that they may not be very effective. For self-help, it is important to learn to monitor your own behaviour, and to use self-reinforcement (maybe that half day at the golf course, or just sitting in the sun with a book!). Initially the farmer must recognize the problem/s. This is where benchmarks come into play, and also keeping records of what happens, and noting your approaches to various problems. Often just the exercise of keeping good records starts the farmer thinking about the situation and consequently considering ways to improve. Perhaps, for example, a review of a farmer's diaries makes it clear he is shifting stock on a regular weekly basis. It is easier to stick to a simple routine rather than move stock from area to area based on feed and growth predictions. Having noted this habit, maybe the farmer will then consider whether an alternative approach would be better.

Usually a farmer will need to practise taking note of his feelings and processes. This is where self-honesty and openness is important, something only the farmer will know occurs, though sometimes a spouse can help in the process from their knowledge of a person gained over many years of observation. Effectively a farmer needs to conduct a personal SWOT analysis:

- strengths;
- weaknesses;
- opportunities; and
- threats.

The opportunities relate to training and improvement opportunities, and the threats to distracting situations. Sorting strengths and weaknesses is where the main effort must go.

And with self-help, it is possible a mentor can help develop this skill through being a simple facilitator encouraging self-analysis and efforts to change thoughts and processes. A mentor in this role must be totally non-judgemental and empathetic as well as encouraging the person to come up with his or her own suggestions through supportive prompting. Of all approaches, this one requires the farmer to be convinced that self-analysis and change are highly desirable. They need to have considerable self-control, or develop it.

At the other end of the scale, group approaches are particularly useful where a farmer has problems relating to other people. The farmer can see the reaction of others as his ideas and comments are expressed, particularly in role play situations. Groups are also a good approach where there are a number of people with the same 'problem'. For example, for farmers who have an inappropriately low 'locus of control', sharing their views and ideas for improvement will be a shared and reinforcing approach. To initiate this, of course, someone has to conduct a test and organize a meeting. Perhaps a professional consultant might do this for a subsection of the farmers who have contracted him to help.

In summary, to change attitudes and inappropriate tacit approaches, systems must be set up that provide interpersonal relationships that exhibit trust and understanding. The system must then provide reassurance and support providing a belief that positive change is indeed possible and real. It must also provide an avenue by which problems that worry people can be shared making them seem less serious and damaging. On the other side of the coin, the process should provide reinforcement for the management skills that are currently good, and similarly of improvement as it evolves.

Providing an understanding of the source of any incorrect management skill and procedure is also usually beneficial. The knowledge that change is possible is also obviously extremely valuable, and there is certainly evidence that farmers improve given the desire and right support. Visualizing what might be possible is always a way to set the goal posts, and records reviewing where a farmer is in the stage of reaching the 'goal posts' is part of the necessary feedback (Aspinal and Taylor, 1997; Taylor et al., 1998).

Where mentoring takes place, this is often referred to as a 'participatory' approach as the farmer is participating in contrast to being lectured at. Some Dutch evidence in which farmers were mentored in small groups shows just how effective this technique can be. Wolf et al. (2004, p. 205) conclude: 'The identification of weaknesses is an essential element to motivate farmers to improve their entrepreneurial qualities' and '[t]he group learning process contributes to the improvement of individual entrepreneurship, facilitating farmers to learn from colleagues and to become more conscious about personal strengths and weaknesses. The comparison of individual results and the discussion with colleagues are very useful methods to create a group learning process' (p. 206).

Case farmer views

As he frequently pointed out, Margrave always enjoyed enormously trying out new technology and ventures. This adventurousness fed off professional agriculturists. Margrave discovered two particularly compatible consultants, one employed by the Government-funded Ministry of Agriculture and Fisheries, and the other in private practice. These people became his mentors with whom he shared most aspects of his operation, and they in turn came to him for comments on ideas and technologies they were keen on. This was a mutually beneficial arrangement which gave Margrave considerable confidence, assistance and support. One suspects that his high intellect and forward-looking attitude went beyond many farmers and so he bypassed nearby colleagues to provide the kind of mentoring he looked for, and found it in the professionals who happened to live and work in his area. Margrave comments: 'It was a two way relationship with all of us bouncing off each other with all sorts of ideas.'

Margrave believes intuition is an important aspect to his decision making and he does have considerable confidence in the decisions that seem to just appear. He comments: 'I'm quite an intuitive farmer. I guess this arises from the intelligent use of experience. I've been doing all that kind of thing since I was a little fellow so I must have learnt something.' To Margrave, reflection on all his experiences to come to a conclusion on the correct approach is important and adds to the development of reliable and correct intuition. Careful attention to appropriate analyses over the years helps develop this intuition, as does a healthy review of the current conditions to ensure that the intuitive response is really applicable. His story of the potential holiday home comes to mind; his strong intuition led him to a logical conclusion.

When asked if he thought his management style could be changed for the better, Margrave was quick with 'probably not greatly', but on reflection he was not so sure. He ventured it was possible to 'change round the margins' and noted: 'I gained in confidence with encouragement and support and consequently improved. If feeling confident you make the right decisions'.

Margrave also found mentor groups valuable and obtained valuable thoughts on current issues. He noted they were 'a great sounding board, a great medium for all sorts of issues. We trusted each other and treated the discussions with total confidence'. The discussions did not distinguish between farm and family and so our total lives were under scrutiny. Margrave also found his involvement with a national group designed to explore all issues facing his industry valuable from a technical viewpoint. These meetings invite leading scientists and agriculturists to present their ideas and engender critical discussions and eventual conclusions. But this forum of farmers was largely divorced from each other's immediate problems.

But what was more intimately involved in Margrave's decision problems was his use of computer software to analyse his data. 'My consultant started me on using the computer, and I learnt reasonably quickly.' He admitted he started by playing computer games, but 'now I have all the skills and get totally bored with them…can't be bothered. I need something more challenging and have found this in sorting out farm issues.' Margrave is involved in planning all the feed use on the farm and constantly updates his estimates and calculations forecasting supply and demand. He also has computerized farm maps for planning activities and fertilizer use. Spreadsheets and word processing are now second nature. But the important thing for Margrave is that 'half the benefit of using the software is the need to process the information which then passes through my mind…this active involvement starts the subconscious going and stimulates the intuition…the answers become obvious'.

Hank also finds his computer a very important part of his management 'team' with its constant use on cash flow updates, annual budgets, feed planning as well as its use to access the dairy industry central records. These provide monthly updates on individual cow production as well as the ongoing totals. Another important, but passive, member of his team is his diary. It gets used to record goals, job lists, and achievements leading to continual job list updates, and their re-prioritization. The diary gets attended to the first thing each day. 'I get muddled if I don't write it all down, and I certainly don't like feeling out of control.' Given the size of Hank's operation one can understand how he feels. The diary provides order, and its use is a chance to think through all the issues.

Hank has made, and does make, constant use of mentors and groups. He used to belong to a reasonably sized discussion group, but comments: 'I found this group useful when I first started, but now find in such a big group too much time is wasted discussing practical issues like how much pasture is in a field.

I now find I want more specific information relating to my situation. The meetings ended up being more of a social outing, which is, of course, very important to new people in the area'.

What Hank now finds more relevant and useful is a small group of friends who regularly review each other's actions and decisions. Hank also finds his banker is a great sounding board with whom he constantly talks especially after sending updated budgets (which occurs on a regular basis). These relationships depend very much on the respect each has for the other. Hank's accountant similarly contributes, particularly, over appropriate business structures and related taxation matters.

Hank is also a great believer in using professional consultants as a lynchpin mentor. 'I use consultants all the time for they know my situation, and also know "me". The first consultant I had was just great for the basic stuff such as feeding systems and calculations, but I improved my skills here, and no longer needed this kind of support. I have now moved onto a different kind of consultant with an emphasis on the business side of my operations. The trouble is I'm a strong character and consequently require someone who will stand up to me and provide challenges. Together we get it mainly right.'

Hank believes some consultants just say what the farmer wants to hear, and tend to be 'pollinators', simply passing from one farmer to the other the best ideas: 'This is a waste of time for experienced farmers.' He concludes: 'The personality mix is very important, and the guy has to have credibility in the eyes of the farmer client.'

Professional consultant's views

Bruce has strong views on the role of consultants in helping farmers improve their management. Above all Bruce believes he should challenge his farmer clients, so that they can assess their own situation and come to conclusions on how to improve. Of course this 'challenge' is given structure by the consultant with a view to lead the farmer along logical paths.

Bruce also notes that the consultant himself must always be learning, and the methods are similar to what a farmer should be doing. The consultant must always explore new situations and gain experience accordingly as well as constantly reading and investigating new ideas. Part of the learning is having extensive networks where others can challenge and provide new insights. As there is always a danger that a consultant can become isolated and stagnated, it is important to belong to professional organizations and societies.

Some consultants call themselves 'business coaches' and encourage the farmers to talk about their plans and programmes, so that they can be challenged over what they are doing. This process helps the farmer develop their capacity to acquire improved skills. They learn to:

- identify;
- observe;
- assess alternatives; and then
- decide and review.

In the end, consultants, Bruce believes, must be positive people: 'Here are the opportunities, let's try this one, it is very exciting.' Some farmers improve with this support and the consultant can 'let them operate on their own, but others always need their hands held'. If a consultant 'puts enough fires under a client the spark to start up something new may be created'. The consultant must convince the farmer new approaches are possible, and in the process share some of the risk and provide confidence. Sometimes it is necessary to 'show a farmer that he can't afford not to change' and this can change his risk attitude, 'you are going to regret not doing this'.

One of the advantages a consultant can bring is a new perspective from an outsider's view, especially if the consultant has extensive networks. 'Consultants play a big role in being agents for change.' Bruce comments that farmers can become very introspective with the lack of outside stimulation. He also notes that the relationship between the consultant and the farmer is critical. A trust and respect must be developed, and this will only occur if indeed this trust and respect is deserved.

Any consultant must really listen to the farmer and determine his beliefs and processes. In this respect, it is important to match the farmer and the consultant personality wise: 'some match ups just do not work' (Hank also stressed this point). Sometimes initially the consultant must carefully guide the farmer, but for real progress this relationship must progress. The consultant cannot take all the risk for the farmer, and, besides, doing this can be very risky for the consultant himself. Success is dependent on the outcome of many risk factors over which the consultant has no more control than the farmer.

As has been stressed, one of the first steps is for the farmer to be very clear on his objectives. The consultant, Bruce believes, can make a real contribution in this regard challenging the farmer's professed objectives and goals. Bruce comments: 'It often takes time to drill down to the real goals.' By way of an example, Bruce noted that the professed maximum profit motive often stems from a desire to be better than the neighbours. The consultant needs to work away at sorting out these real motives, and where they originate.

Prof has similar views about the farmer's objectives, and comments: 'It is absolutely no use asking a farmer what his objectives are. He'll just trot out what you want to hear.' A consultant must watch, listen and observe to really find out the real objectives. The consultant must work out what the farmer is doing and why through observing incidences and events.

In a similar vein to Bruce, Prof notes that farmers 'need measures of success...its all about how you feel about yourself'. Thus, 'profit maximization is only important to the extent it is a measure of success' with farmers often having quite basic objectives for which they strive. Some farmers can be called 'developers' and their objective is to enhance and grow their assets, and others are simply 'farmers' trying to get high levels of production which may, or may not, lead to profit maximization.

In improving a farmer's ability, Prof believes you must first understand his personality, skills and weaknesses. Most would agree with this approach. Another important factor is understanding just what it is the farmer wants from a consultant, as, after all, the farmer is paying the bill and should 'call the shots'. One survey of a small group of consultants revealed that 37% of the farmers wanted help with technical factors (how, what, etc.), 20% wanted business coaching and motivation (help with might be called the soft skills), and 43% wanted a mixture of technical and coaching help. These different demands means there are 'horses for courses', so care must be taken to ensure that the client and consultant combination works.

Prof also believes most farmers learn by 'seeing and doing', so the ability to see what other farmers are doing is very important. Often a farmer will try something in a small way as part of 'seeing and doing'. As noted earlier, Prof commented that the ability to find out what other farmers are doing is very important and in this sense a consultant can foster this 'across the fence' observation. Prof stressed that some farmers have useful strategies to get on to other farms. One farmer who sold stud bulls always visited his clients each year ostensibly to see how the bulls were helping the client, but in practice it was to spend time on other farms to see what the farmer was doing, and why. Another farmer was known by his children as 'sidetracker' for whenever they went on a holiday, they shot off on to side roads so that dad could see and talk to other farmers.

Prof considers role models are very important. Thus, farmer competitions play an important role as the successful farmers are available for other farmers to model themselves on. In many modern agricultural industries, these competitions abound: the best manager of the year, the best dairy businessman of the year, the best young farmer of the year, and so on. Field days also play a part in that the winners are usually required to open their farm and systems up to others.

Prof also related that one of his students studied the features of successful farmers. Two important attributes were the time spent looking over the fence at other farmers, including attending field days on the competition winner's farms, and the time spent reading each day. These are both means of obtaining ideas. A consultant can encourage and stimulate these activities including helping the farmer critique what he sees and reads about.

'Leading farmers are very important to a community.' Prof believes in the trickle down theory which considers that improved systems and methods will filter through the farming population from the leading farmers. If the community of farmers is open and considerable sharing occurs, this can indeed operate. Thus, the importance of role models talked about earlier. But this approach largely involves practical farming systems in contrast to changing personal attributes. This is where an individual farmer approach is necessary. Prof reckons, however, it is up to the farmer to change, for you can make a farmer aware of his strengths and weaknesses, but you cannot impose improvement programmes. Some farmers have the basic attributes, but others just cannot be made into, for example, an entrepreneur. And when talking about families, Prof commented consultants must be cautious. They are not trained counsellors, and besides any move beyond the farmer himself must be stimulated by the farmer's employer, the client is the 'king' and should control what happens.

The final word Prof had about change techniques concerned discussion groups. One of his first jobs as an adviser was organizing these groups in which farmers were brought together to discuss the topics of the day, and often to inspect each other's farms leading to discussions on what could be changed. These groups are extremely valuable, but must be reinvigorated from time to time as they have their own natural life cycle leading to stagnation. The topics dry up, and the farmer's gain little more having already acquired the skills on offer. The leader of the group is critical to its success and must have an extrovert tendency to always create challenge and excitement. And to be successful, the members of any group must have the same learning needs, otherwise they get bored and drift away.

Prof's contribution to this discussion provides a different perspective emphasizing industry approaches rather than individual farmer-focused systems. From a farmer organizational view, and a government view, these are valuable insights.

Concluding Comments

Success depends on an individual's abilities, though good luck in this risky and uncertain world helps. However, as it is not possible for an individual to make major changes to the environment in which he works, helping a person achieve success involves helping them improve their skills best suited to the environment. Achieving this requires understanding the person, and understanding how farmers acquire the level of skills they currently have in addition to understanding how changes can be made. Changing simple factual knowledge is relatively straightforward, but changing skills stemming from the inner human workings is more difficult. This last chapter has considered approaches to this problem, whereas previous chapters have provided information on the factors giving rise to a farmer's skill level, and on the skills that a farmer must be good at to succeed.

In improving managerial skill it is important for a farmer, or his professional helpers, to have some kind of plan. To do this, an assessment of the types of skills that need improving must be conducted leading to a list of skills and processes to be used in achieving improvement. This list should be put into priority order for it is easier to work on one skill at a time. As the plan moves on, improved skills must be constantly practised and reinforced at the same time as starting on improvement of a new skill. Invariably knowledge and understanding deepens with further practice as seldom does one session create an expert.

To assess which skills need improving any tests available can be used (and several have been supplied in this book), and similarly the observations of a professional who has worked with the farmer for some time are helpful. The farmer's family may also have insights into what is required provided the farmer is willing to have their involvement. Similarly employees and contractors may have useful ideas again provided the farmer is happy to receive what may be seen as criticisms.

In deciding on what aspects of management to improve, the farmer himself may well have ideas that should also be included in the list. For the mentor/s, it is helpful to inspect the farmer's farm, and the records and accounts to better judge the nature of the decision problems. The next step is to start the improvement process using the approaches outlined. The meetings and discussions then continue until it is deemed that the list of problems has been sorted, and the lessons absorbed and understood. The farmer himself may well be a good judge of this.

One objective is to improve the farmer's tacit knowledge. Decisions and methods then become automatic and do not require careful and time-consuming analysis. However, the tacit knowledge must be continually assessed, and updated if the conditions change sufficiently. An example might be where there is a fundamental shift in the markets for a group of products, or perhaps the tax laws change markedly, or there might be a significant technological change.

Clearly many farmers who could benefit from skill improvement do not make efforts in this direction. Where their financial situation provides an adequate living, in contrast to an appropriate return on the investment, they may well be content. Others may wish to improve, but are fearful of exposing their situation to colleagues and professionals. In each of these cases, it is only possible to point out what is possible, and what they might be missing out on in the efficiency stakes. If they have contact with cases where improvement has occurred, and the person makes the benefits clear, that will encourage participation. The old saying is 'you can take the horse to water, but you can't make it drink'. So true.

References

Agrawal, R.C. and Heady, E.O. (1972) *Operations Research Methods for Agricultural Decisions*. Iowa State University Press, Ames, Iowa.

Ajzen, I. (1991) The theory of planned behaviour. *Organisational Behavior and Human Decision Processes* 50, 179–211.

Anderson, J., Dill, J. and Hardaker, B. (1977) *Agricultural Decision Analysis*. Iowa State University Press, Ames, Iowa.

Anderson, J.R., Dillon, J.L. & Hardaker, B. (1977) *Agricultural Decision Analysis*. Iowa State University Press, Ames, Iowa.

Ashenfelter, O. and Rouse, C. (1998) Income, schooling, and ability: evidence from a new sample of identical twins. *The Quarterly Journal of Economics* 253–284.

Ashenfelter, O. and Rouse, C. (1998) Income, schooling, and ability: evidence from a new sample of identical twins. *The Quarterly Journal of Economics* 113(1), 253–284.

Aspinal, L.G. and Taylor, S.E. (1997) A stitch in time: self regulation and procedure coping. *Psychological Bulletin* 121(3), 417–436.

Atkinson, R.L., Atkinson, R.C., Smith, E.E., Bem, D.J. and Nolen-Hoeksema, S. (1999) *Hilgard's Introduction to Psychology*. Harcourt Brace, Fort Worth, Texas.

Atkinson, R.L., Atkinson, R.C., Smith, E.E., Bem, D.J. and Nolen-Hoeksema, S. (1996) *Hilgard's Introduction to Psychology*, 12th edn. Harcourt Brace, Fort Worth, Texas.

Atkinsonm, R.L., Atkinson, R.C., Smith, E.E., Bem, D.J. and Nolen-Hoeksema, S. (1996) *Hilgard's Introduction to Psychology*. Harcourt Brace, Fort Worth, Texas.

Austin, E.J., Willock, J., Deary, I.J., Gibson, G.J., Dent, J.B., Edwards-Jones, G., Morgan, O., Grieve, R. and Sutherland, A. (1998) Empirical models of farmer behaviour using psychological, social, and economic variables – part 1: linear modelling. *Agricultural Systems* 58, 203–224.

Beaudry, P. and Francois, P. (2005) Managerial skills and the theory of economic development. NBER Working Paper, Department of Economics, University of British Columbia, Vancouver.

Beedell, J. and Rehman, T. (2000) Using social-psychology models to understand farmers' conservation behaviour. *Journal of Rural Studies* 16, 117–127.

Bentler, P.M. and Newcomb, M.D. (1978) Longitudinal study of marital success and failure. *Journal of Consulting and Clinical Psychology* 46(5), 1053–1070.

Bokemeier, J. and Garkovich, L. (1987) Assessing the influence of farm women's self identity on task allocation and decision making. *Rural Sociology* 52(1), 13–36.

Bowles, S., Gintis, H. and Osborne, M. (2001) The determinants of earning: A behavioral approach. *Journal of Economic Literature* 39, 1137–1176.

Cameron, D. and Chamala, S. (2002). Measuring impacts of an holistic farm business management training program. Proceedings of the 13th International IFMA congress of Farm Management.

Case, H.C.M. and Johnson, P.E. (1953) *Principles of Farm Management*. J.B. Lippincott, Chicago, Illinois, 446 pp.

Cattell, R.B., Eber, H.W. and Tatsuoka, M.M. (1970) *Handbook for the Sixteen Personality Factor Questionnaire*. Institute for Personality and Ability Testing, Champaign, Illinois.

Caughlin, J.P., Huston, T.L. and Houts, R.M. (2000) How does personality matter in marriage? An examination of trait anxiety, interpersonal negativity, and marital satisfaction. *Journal of Personality and Social Psychology* 78(2), 326–336.

Costa, P.T. and McCrae, R.R. (1992) *NEO-PI-R Professional Manual*. Psychological Assessment Resources, Odessa, Florida.

Curland, S.R. and Fawcett, S.L. (2001) Using simulation and gaming to develop financial skills in undergraduates. *International Journal of Contemporary Hospitality Management* 13(3), 116–121.

de Lauwere, C.C. (2005) The role of agricultural entrepreneurship in Dutch agriculture today. *Agricultural Economics* 33(2), 229–238.

Dhungana, B.R. (2000) *Measuring and Explaining Economic Inefficiency of Nepalese Rice Farmers*. Masters thesis, Lincoln University, Canterbury, New Zealan.

Enos, M.D., Kehrhahn, M.T. and Bell, A. (2003) Informal learning and the transfer of learning: how managers develop proficiency. *Human Resource Development Quarterly* 14(4), 369–387.

Ewell, P.T. (1997) *Organizing for Learning: a Point of Entry*. Available at: www.intime.uni.edu/model/learning/learn_summary.html. Accessed 21/8/2009

Feldman, S. and Walsh, R. (1995) Feminist knowledge claims. Local knowledge and gender division of agricultural labor. *Rural Sociology* 60, 23–43.

Gardner, H. (1993) *Multiple Intelligence: The Theory and Practice*. Basic Books, New York.

Garforth, C. and Rehman, T. (2005) Review of literature on measuring farmers' values, goals and objectives. Project report no. 2 in the 'Research to understand and model the behaviour and motivations of farmers responding to policy changes (England)'. School of Agriculture, Policy and Development, The University of Reading, Reading, UK.

Gasson, R. (1973) Goals and values of farmers. *Journal of Agricultural Economics* 24(3), 521–542.

Hanson, D.G., Parsons, R.L., Chess, W.J., and Balliet, K.L. (2002) Farm production analysis training for small farmers. *Journal of Extension* 40(4), 1–8.

Higgs, M. (2001) Is there a relationship between the Myers-Briggs type indicator and emotional intelligence? *Journal of Managerial Psychology*, 16(7), 509–533.

Hobbs, D.J., Beal, G.M. and Bohlen, J.M. (1964) The relation of farm operator values and attitudes to their economic performance. Rural Sociology Report No. 33, Department of Economics and Sociology, Iowa State University, Iowa.

Howard, W.H., Brinkman, G.L. and Lambert, R. (1997) Thinking styles and financial characteristics of selected Canadian farm managers. *Canadian Journal of Agricultural Economics* 45, 39–49.

Jackson-Smith, D., Trechter, D. and Splett, N. (2004) The contribution to financial management training and knowledge in dairy farm financial performance. *Review of Agricultural Economics* 26(1), 132–147.

Johnson, G.L., Halter, A.N., Jensen, H.R. and Thomas, D.W. (eds) (1961) *A Study of Managerial Processes of Midwestern Farmers*. Iowa State University Press, Ames, 221pp.

Jose, H.D. and Crumly, J.A. (1993) Psychological type of farm/ranch operators: relationship to financial measures. *Review of Agricultural Economics* 15, 121–132.

Judge, T.A. and Ilies, R. (2002) Relationship of personality to performance motivation: a meta-analytic review. *Journal of Applied Psychology* 87(4), 797–807.

Judge, T.A. and Ilies, R. (2002) Relationship of personality to performance motivation: a meta analytic review. *Journal of Applied Psychology* 87(4), 797–807.

Kaine, G., Sandall, J. and Bewsell, D. (2004) *Personality and Strategy in Agriculture: Proceedings of the 20th Annual Conference*. AIAEE, Dublin. pp. 790–801.

Kanfer, R. and Ackerman, P. (2000) Individual differences in work motivation: further explorations of a trait framework. *Applied Psychology* 49(3), 470–482.

Kayes, D.C. (2002) Experiential and its critics. Preserving the role of experience in management learning and education. *Academy of Management Learning and Education* 1(2), 137–149.

Keirsey, D. (1998) *Please Understand Me II*. Prometheus Book Co, California.

Kelly, E.L. and Conley, J.J. (1987) Personality and compatibility: a prospective analysis of marital stability and marital satisfaction. *Journal of Personality and Social Psychology* 52(1), 27–40.

Kelly, G.A. (1992 reprint). *The Psychology of Personal Constructs*. Vol. 1. Routledge, UK.

Kilpatrick, S. (1997) Education and training: impacts on profitability in agriculture. *Australian and New Zealand Journal of Vocational Education Research* 5(2), 11–36.

Kilpatrick, S. (1999) Education and training: impacts on farm management practice. Centre for Research and Learning in Regional Australia, University of Tasmania, Launceston, Tasmania.

Kim, J. and Zepeda, L. (2004) When the work is never done: time allocation in US family farm households. *Feminist Economist* 10(1), 115–139.

Kolb, D.A. (1984) *Organizational Psychology*, 4th edn. Prentice-Hall, New Jersey.

Kolb, D.A. (2005) *Kolb Learning Style Inventory 3.1*. Haygroup, Boston, Massachusetts.

Kolb, D.A., Rubin, I.M. and McIntyre, J. (1974). *Organizational Psychology. A Book of Readings*. Prentice-Hall, Englewood, New Jersey.

Matthews, G. and Deary, I.J. (1998) *Personality Traits*. Cambridge University Press, Cambridge.

McCall, M., Lombardo, M. and Morrison, A. (1988) *The Lessons of Experience: How Successful Executives Develop on the Job*. Lexington Press, Lexington, Massachusetts.

Menz, K.M. and Longworth, J.W. (1976) An integrated approach to farm management education. *American Journal of Agricultural Economics* 58(3), 551–556.

Muggen, G. (1969) Human factors in farm management: a review of the literature. *World Agricultural Economics and Rural Sociology Abstracts* 11, 1–11.

Nielsen, A.H. (1974) A management game for use in teaching farm management. *European Review of Agricultural Economics* 2(3), 293–306.

Nuthall, G. (2007) *The Hidden Lives of Learners*. NCER Press, Wellington, New Zealand.

Nuthall, G.A. and Alton-Lee, A. (1993) Predicting learning from student experience of teaching. A theory of student knowledge construction in classrooms. *American Educational Research Journal* 30(4), 799–840.

Nuthall, P. and Benbow, C. (1999) *Computer System Uptake and Use on New Zealand Farms*. Research Report 99/01, Farm & Horticultural Management Group, Lincoln University, Christchurch, New Zealand.

Nuthall, P.L. (2002) Managerial competencies in primary production: the view of a sample of New Zealand farmers. FHMG Research Report 02/2002, Lincoln University (ISSN 1174–8796).

Nuthall, P.L. (2006) Determining the important management skill competencies: the case of family farm business in New Zealand. *Agricultural Systems* 88, 429–450.

Nuthall, P.L. (2009) Modelling the origins of managerial ability in agricultural production. *Australian Journal of Agricultural and Resource Economics*. 53, 413–436.

Ohlmer, B., Olson, K. and Brehmer, B. (1998) Understanding farmers' decision making processes and improving managerial assistance. *Agricultural Economics* 18, 273–290.

Psacharopoulos, G. and Patrinos, H.A. (2004) Returns to investment in education: a further update. *Education Economics* 12(2), 111–134.

Rickson, S.T. and Daniels, P.L. (1999) Rural women and decision making: women's role in resource management during rural restructuring. *Rural Sociology* 64(2), 234–250.

Rougour, C.W., Tripp, G., Huirne, R.B.M. and Renkema, J.A. (1998) How to define and study farmers' management capacity: theory and use in agricultural economics. *Agricultural Economics* 18, 261–272.

Samobodo, L. and Nuthall, P.L. (2009) *The Decision Making Processes of Semi-Commercial Farmers. A Case Study of Technology Adoption in Indonesia.* VDM, Saarbrucken, Germany.

Sawer, B. (1974) The role of the wife in farm decisions. Rural Sociology Monograph No. 5 104pp.

Solano, C., Leon, H., Perez, E., Tole, L., Fawcett, R.H. and Herrero, M. (2006) Using farmer decision making profiles and managerial capacity as predictors of farm management performance in Costa Rican dairy farmers. *Agricultural Systems* 88, 395–428.

Sternberg, R.J. (1995) *In Search of the Human Mind.* Harcourt Brace, Fort Worth, Texas.

Sternberg, R.J., Wagner, R.K., Williams, W.M. and Horvath, J.A. (1995) Testing common sense. *The American Psychologist* 50(11), 912–927.

Stewart, W.H., Watson, W.E., Carland, J.C. and Carland, J.W. (1998) A proclivity for entrepreneurship: a comparison of entrepreneurs, small business owners, and corporate managers. *Journal of Business Venturing* 14(2), 189–214.

Summer, D.A. and Lieby, J.D. (1987) An econometric analysis of the effect of human capital on size and growth among dairy farms. *American Journal of Agricultural Economics* 69, 465–470.

Taylor, S.E. and Armor, D.A. (1996) Positive illusions and coping with adversity. *Journal of Personality* 64(4), 873–898.

Taylor, S.E., Pham, L.B., Rivkin, I.D. and Armor, D.A. (1998) Harnessing the imagination. Mental stimulation, self regulation and coping. *American Psychologist* 53(4), 429–439.

Thomas, H.V., Lewis, G., Thomas, D.Rh., Salmon, R.L., Chalmers, R.M., Coleman, T.J., Kench, S.M., Morgan-Capner, P., Meadows, D., Sillis, M. and Softley, P. (2003) Mental health of British farmers. *Occupational and Environmental Medicine* 60, 181–186.

Tripp, G., Thijssen, G.J., Renkema, J.A. and Huirne, R.B.M. (2000) Measuring managerial efficiency: the case of commercial greenhouse growers. *Agricultural Economics* 27, 172–181.

Warren, R.D., White, J.K. and Fuller, W.A. (1974) An errors in variables analysis of managerial role performance. *Journal of American Statistical Association* 69, 886–893.

Willock, J., Deary, I.J., McGregor, M.M., Sutherland, A., Edwards-Jones, G., Morgan, O., Dent, B., Grieve, R., Gibson, G. and Austin, E. (1999) Farmers' attitudes, objectives, behaviours, and personality traits: The Edinburgh study of decision making on farms. *Journal of Vocational Behaviour* 54, 5–36.

Wilson, P., Hadley, D., Ramsden, S. and Kaltsas, I. (1998) Measuring and explaining technical efficiency in UK potato production. *Journal of Agricultural Economics* 49, 294–305.

Wolf de, P.L., Schoorlemmer, H.B., Smit, A.B. and de Lauwere, C.C. (2004) Analysis and development of entrepreneurship in agriculture. In: Bokeman, W. (ed.) Proceedings of the XVth International Symposium on Horticultural Economics and Management. *ISHS Acta Horticulturae* 655.

Young, A.J. and Walters, J.L. (2002) Relationship between DHI production values and Myers-Briggs type indicator as a measure of management ability. *Journal of Dairy Science* 85, 2046–2052.

Zeidner, M., Matthews, G. and Roberts, R.D. (2004) Emotional intelligence in the workplace. *Applied Psychology* 53(3), 371–390.

Zuckerman, M., Joireman, J., Kraft, M. and Kuhlman, D.M. (1999) Where do motivational and emotional traits fit within the three factor models of personality? *Personality and Individual Differences* 26(3), 487–504.

Index